21 Fortschritte der chemischen Forschung
Topics in Current Chemistry

Organic
Electrochemistry

Springer-Verlag
Berlin Heidelberg GmbH **1971**

ISBN 978-3-540-05463-4   ISBN 978-3-540-36652-2 (eBook)
DOI 10.1007/978-3-540-36652-2

Originally published by Springer-Verlag Berlin Heidelberg New York in 1971

Library of Congress Catalog Card Number 51-5497.

Contents

ORGANIC ELECTROCHEMISTRY

Prof. Dr. L. Eberson

University of Lund, Chemical Center, Lund, Sweden

Doz. Dr. H. Schäfer

University of Göttingen, Organisch-Chemisches Institut, Göttingen, Germany

Contents

Introduction

The standard definitions of oxidation and reduction rest on the concept of an electron transfer between two reagents. This is widely recognized in inorganic redox reaction mechanisms [1,2] but has received far less attention in organic chemistry [3]. Maybe this is one important reason why electrochemical synthetic procedures, based on electron transfer between a metal electrode (normally inert) and a substrate by aid of the electric current, are seldom considered as methods of choice for performing oxidations or reductions of organic compounds. The almost complete lack of mention of organic electrode reactions in textbooks and experimental courses bears witness to the neglect of what might often be superior procedures. This is in spite of the fact that electrolytic reactions nowadays are no more complicated in experimental design than catalytic hydrogenations or photochemical reactions.

While general laboratory practice of organic chemistry still makes very little use of electrochemical methods, the study of organic electrode processes has nevertheless received considerable attention during recent years, resulting in a better understanding of which factors are important in the design of electrochemical syntheses. To be fair, it must be admitted that mechanistic organic electrochemistry is only in its infancy at present, and that the problem of formulating mechanisms in this field is of enormous complexity due to the heterogeneity of the electrode process. Organic chemists, prone to write mechanisms in simple structural terms (and this approach has indeed been very successful in homogeneous solution chemistry), have yet to come to an agreement with physical chemists as to whether this is necessary or at all possible in electrochemistry, whereas physical chemists, sometimes prone to neglect the structural aspects, will have to convince organic chemists about the necessity of including their viewpoints about electrode kinetics. At present, many issues are still controversial due to this difference in emphasis, but there is a definite trend towards a common view of the mechanistic problem.

The wide range of activities in electroorganic chemistry is reflected in a large number of reviews and textbooks. Synthetic and experimental aspects are stressed in the older literature, of which the textbooks by Fichter [4], Brockmann [5] and Allen [6] are outstanding examples as well as the exhaustive reviews written by Swann [7] and Meites [8]. Of more recent origin is a review on the cathodic reduction of organic compounds [9], covering the literature between 1940 and 1960. More mechanistically oriented — in the sense of organic chemistry — are recent reviews on electrochemical oxidation of organic compounds [10,11], reactions of cathodically generated radicals and anions [12], controlled potential electrolysis in organic chemistry [13], electrochemical reactions of carboxylic acids [14], mechanisms of organic polarography [15], anodic substitution reactions [16], cathodic coupling [388], electrolytic cyclization [432], cathodic reduction of the C=N linkage [155b], free radical reactions in the electrolysis of organic compounds [17], and electroorganic chemistry [18,18a,18b].

The physico-chemical approach to organic electrode processes has been summarized in a textbook by Conway [19] and further elaborated in reviews on the anodic oxidation of organic compounds [20] and on electrode kinetic aspects of the Kolbe reaction [21]. The series of monographs edited by Delahay and Tobias [22] and Bard [23] also contain many articles of great interest in this connection.

The use of polarographic methods for the study of substituent effects in organic molecules has been thoroughly treated by Zuman [24]. Voltammetric methods and their use for the study of organic solid electrode processes are excellently treated in a recent textbook by Adams [25]. Of great value are the bibliographies of electroorganic reactions published by Swann [26], as well as those of the electroorganic patent literature up to 1960 by the same author [27]. Finally, a number of articles give brief accounts of different aspects of the field [28-34]

This book aims at a presentation of electrochemical methods of particular interest to the organic chemist who is looking for alternative synthetic procedures. Consequently, the emphasis will be laid on

1. those simple techniques which can aid in the design of an electrochemical synthetic method,

2. experimental factors and their influence,

3. experimental procedures for carrying out electrolyses,

4. electrolytic reactions which have few or no counterparts in ordinary laboratory practice (thus, the most obvious and many times very useful application of converting a functional group into another by electrolysis will be treated only briefly), and

5. electrolytic reactions which do have counterparts in ordinary laboratory practice but are much simpler to perform electrochemically on a reasonably large scale (e.g., cathodic reductions instead of metal reductions).

2. Classification of Electro-Organic Reactions

Electroorganic reactions may be conveniently classified on the basis of products formed into the following categories:

1. Conversion reactions (one functional group into another one),
2. Substitution reactions,
3. Addition reactions,
4. Elimination reactions,
5. Coupling reactions,
6. Cleavage reactions,
7. Electron transfer reactions (with formation of stable radical ions or ions).

All of these reactions types can be effected at the cathode or the anode. However, since organic electrode processes often occur *via* a blend of radical and ionic mechanisms, it should be stressed that the simplicity of the classification scheme above does not imply that the electrochemical oxidation or reduction of a given substrate results in only one type of reaction; on the contrary, many processes give products emanating from two or more of the reaction types above, and it may often involve a lot of experimental work to find optimum conditions for a (desired) reaction to occur, if it is at all possible.

In the following treatment, electroorganic reactions will be classified after the type of reaction leading either to the predominant product, or, if this is of a rather trivial nature, to a minor product of particular interest. In many of these cases, few efforts have been made to increase the yields of such products of interest by changing experimental variables, so there is ample room for systematic work to accomplish this goal. As an example of what can be achieved by simply changing the electrode material, anodic oxidation of ethyl vinyl ether in methanol-KOH at a *platinum* electrode leads to formation of an addition product *(1)* and a coupling-addition product *(2)* with the former one predominating [35];

$$
\text{EtO-CH=CH}_2 \xrightarrow[\text{KOH, Pt}]{\text{MeOH}} \overset{\overset{\text{OMe}}{|}}{\text{EtO-CH-CH}_2\text{OMe}} + \overset{\overset{\text{OMe}}{|}}{\text{EtO-CH-CH}_2}\text{-CH}_2\overset{\overset{\text{OMe}}{|}}{\text{-CH-OEt}} \tag{1}
$$

$$
\begin{array}{cc}
68\% & 22\% \\
1 & 2
\end{array}
$$

If the reaction is instead carried out at a *graphite* anode [36] the formation of *1* is completely suppressed and *2* is obtained in good yield. It is a sad fact that we cannot at present interpret this difference mechanistically!

A second mode of classification of electrode reactions is based entirely on the electrode mechanism. Here it is necessary to know the number of chemical and electrochemical steps involved and the order between the different steps. By denoting an electrochemical step by an *E* and a chemical one by a *C* and postulating that every electrochemical step involves the transfer of one electron, it is immediately evident that, *e.g.*, an ECEC process consists of:

E: transfer of an electron between substrate and electrode with formation of a radical ion (assuming the substrate to be a neutral molecule containing only paired electrons),

C: reaction between the radical ion and a reagent present to form a second intermediate,

E: electron transfer between this intermediate and the electrode to form a third intermediate, and

C: transformation of the last intermediate into product(s).

As an example, it has been suggested [37] that many anodic substitution reactions of aromatic hydrocarbons are ECEC processes, as shown below for anodic aromatic acetoxylation (*Eqs.* (2-5)):

$$E: \quad Ar\text{-}H \quad \longrightarrow \quad \overset{+\cdot}{Ar\text{-}H} + e^- \tag{2}$$

$$C: \quad \overset{+\cdot}{Ar\text{-}H} + OAc^- \quad \longrightarrow \quad Ar\overset{\cdot}{\diagup}\overset{H}{\diagdown}_{OAc} \tag{3}$$

$$E: \quad \overset{\cdot}{Ar}\overset{H}{\diagup}\diagdown_{OAc} \quad \longrightarrow \quad \overset{+}{Ar}\overset{H}{\diagup}\diagdown_{OAc} + e^- \tag{4}$$

$$C: \quad \overset{+}{Ar}\overset{H}{\diagup}\diagdown_{OAc} + OAc^- \quad \longrightarrow \quad Ar\text{-}OAc + CH_3COOH \tag{5}$$

In order to describe the chemical steps in more detail one can indicate if the attacking reagent acts as a nucleophile (Eq. (3)) or as a base (Eq. (5)) by using the subscripts *N* and *B*, respectively [38]. Thus, in the above example the full description of the mechanism would be EC_NEC_B, a somewhat cumbersome but nevertheless useful denotation.

3. Electrolytic Reactions and Their Use in Organic Synthesis

In this section an attempt will be made at assessing the value of electrolytic methods for synthetic use in organic chemistry. Generally, the use of an electrochemical method should be considered if a known chemical procedure is too laborious, *e.g.,* is of the multi-stage type or involves a tedious work-up procedure, gives too low yields, uses expensive starting materials, and/or cannot easily be run on a large scale. Here electrolytic methods have their main advantages; they often use cheap starting materials, provide short-cuts in multi-stage reaction schemes and give good yields because of the possibility of stepless regulation of the oxidizing or reducing power of the electrode. Moreover, the media employed are normally chemically inert so that further chemical reactions of reactive products can be avoided. Work-up is generally no problem, since there is no need to remove products originating from a chemical oxidant or reductant.

It should also be added that electrochemical reactions often lead to products of unusual stereochemistry as compared to homogeneous reactions (see Sect. 7.2).

Scaling up a reaction in the laboratory often presents unexpected problems, and probably every organic chemist has experience of this. With the recent development of continuously operating cells (Sect. 6.2) a fairly small device should easily handle large quantities of electrolyte and hence provide an attractive alternative to any chemical method. As an example, dissolving metal reductions, notoriously difficult and sometimes expensive to run on a large scale, could preferably be replaced by cathodic reductions in many cases. A recent survey of electrochemistry in Britain has produced a list of reactions, which deserve consideration from the industrial point of view [39].

Needless to say, electrolytic methods should also be considered if no chemical procedure for the preparation of a desired compound exists. Even a superficial knowledge of the principles of electroorganic chemistry, to be outlined in this book, is enough for attacking a synthetic problem from a different perspective. We shall now examine the different reaction types referred to above (Sect. 2.) to see what kinds of transformations are possible. Generally, the denotations given, substitution, addition, etc., do not have quite the same meaning as in conventional organic chemistry, as will become obvious from the following discussion.

3.1. Electrochemical Conversion of a Functional Group into another one

This important reaction type does not differ from its chemical counterpart as far as the result of the reaction is concerned: A functional group is oxidized or reduced to form another functional group, *e.g.*, $-NO_2$ to $-NHOH$ or NH_2, $-CH_2OH$ to $-COOH$ or $-CHO$, $CONH_2$ to $-CH_2NH_2$, etc. Here the main advantages of an electrolytic method appears to be ease of operation and possibility of scale-up. Chemical procedures are normally available and do not present any serious problem in the laboratory. However, one interesting and unusual application is to be found in the preparation of heterocyclic ring systems (Sect. 8.2).

3.2. Electrochemical Substitution

Anodic substitution has some interesting features compared to ordinary substitution reactions. The net reaction is formally between an organic compound and a nucleophile, as in Eq. (6).

$$R\text{-}E + Nu^- \longrightarrow R\text{-}Nu + E^+ + 2\,e^- \qquad (6)$$

$$E \text{ generally} = H$$

In this equation, E normally corresponds to hydrogen but can also be another group, *e.g.* $-OCH_3$. The denotation E is used to show the interplay between nucleophiles and electrophiles in electroorganic reactions. Thus, due to the oxidative nature of the process, a nucleophile can be used as a reagent in an otherwise impossible substitution reaction. As an example of an anodic substitution process, electrochemical cyanation of an aromatic hydrocarbon can be mentioned:

$$Ar\text{-}H + CN^- \longrightarrow Ar\text{-}CN + H^+ + 2\,e^- \qquad (6a)$$

Similarly, electrophiles can be brought into reaction with an organic compound with substitution as a result in a cathodic process. This is schematically depicted in Eq. (7), in which Nu corresponds to a leaving nucleophilic group of some kind. The electrophile could be H^+, CO_2, SO_2, etc.

$$R\text{-}Nu + E^+ + 2\,e^- \longrightarrow R\text{-}E + Nu^- \qquad (7)$$

3.3. Electrochemical Addition

In an anodic addition reaction two molecules of a nucleophile are added across a double bond (Eq. (8)) with loss of two electrons to the anode. This provides an attractive and important route to

$$\text{\Large$\diagdown$C=C$\diagup$} + 2\,Nu^{-} \longrightarrow Nu\text{-}\overset{|}{\underset{|}{C}}\text{-}\overset{|}{\underset{|}{C}}\text{-}Nu + 2\,e^{-} \qquad\qquad (8)$$

bifunctionalization of unsaturated compounds.

Cathodic addition reactions are of great preparative value, since the general scheme (9) of course accommodates the synthetically important reduction of unsaturated compounds:

$$\text{\Large$\diagup$C=C$\diagdown$} + 2\,E^{+} + 2\,e^{-} \longrightarrow E\text{-}\overset{|}{\underset{|}{C}}\text{-}\overset{|}{\underset{|}{C}}\text{-}E \qquad\qquad (9)$$

3.4. Electrochemical Elimination

Formally, the anodic and cathodic elimination reaction is simply the reverse of cathodic and anodic addition, respectively. As we shall see later, the same E and Nu do not always work in the two types of reactions.

3.5. Electrochemical Coupling

The coupling reaction is perhaps the most useful among the different electrochemical reaction types, since it normally has few or no counterparts in conventional laboratory practice. It provides a simple and direct route to bifunctional dimeric compounds from monofunctional monomeric compounds. Anodic coupling processes are formally of two kinds, either a coupling-elimination or a coupling-addition reaction, as shown in Eq. (10) and (11).

$$2\,R\text{-}E \longrightarrow R\text{-}R + 2\,E^{+} + 2\,e^{-} \qquad\qquad (10)$$

$$2\,\text{\Large$\diagdown$C=C$\diagup$} + 2\,Nu^{-} \longrightarrow Nu\text{-}\overset{|}{\underset{|}{C}}\text{-}\overset{|}{\underset{|}{C}}\text{-}\overset{|}{\underset{|}{C}}\text{-}\overset{|}{\underset{|}{C}}\text{-}Nu + 2\,e^{-} \qquad\qquad (11)$$

On the cathodic side, analogous coupling reactions can be realized, as shown in Eq. (12) and (13). The latter process is known under the name of cathodic *hydrodimerization.*

$$2\ R\text{-}Nu + 2\ e^- \longrightarrow R\text{-}R + 2\ Nu^- \tag{12}$$

$$2\ \overset{\diagdown}{\underset{\diagup}{C}}{=}\overset{}{\underset{}{C}}\overset{\diagup}{\diagdown} + 2\ E^+ + 2\ e^- \longrightarrow E\text{-}C\text{-}C\text{-}C\text{-}C\text{-}E \tag{13}$$

3.6. Electrochemical Cleavage

Electrochemical cleavage reactions are not easily accommodated within a general scheme. In many cases they might well be classified as substitution processes; however, the preparative outcome normally would seem to make the reaction conform more closely with a cleavage process.

3.7. Electron Transfer

Anodic and cathodic electron transfer is the elementary process underlying all electrolytic reactions, and is followed by chemical reactions of the intermediates formed in the majority of cases. In cases where the intermediate is stable under the prevailing reaction conditions, anodic or cathodic preparation of radical ions, radicals, or ions is possible. The electrolytic method is here often superior to chemical methods, since work-up is not complicated by the presence of any inorganic redox couple.

4. Methods for the Study of Electro-Organic Reactions

For the organic chemist, product studies in the widest sense, *i.e.,* including stereo-chemical aspects, isotope effects, etc. fall most natural in the study of electro-organic reactions. However, there are also some simple electrochemical techniques which are extremely useful in the design of electrochemical syntheses and can be set up in any laboratory for a modest cost. These methods — which are the ones to be discussed here — include different kinds of voltammetry, controlled poten-tial electrolysis, and coulometry, and give information as to the nature of the elec-tro-active species, the possible nature of intermediates involved and their reactions with reagents present, and the number of electrons involved in the process.

For the investigator who wants to study electrode processes at depth, a num-ber of more physically oriented methods are available, such as double layer capaci-tance measurements [19], rotating disc and ring disc techniques [25], and radio-active tracer methods [40a]. Spectroscopical methods in conjunction with optically transparent electrodes can be used for the study of intermediates [40b], as can also total reflectance spectroscopy [40c].

ESR spectroscopy has found wide-spread use for the detection of radical intermediates in electrode processes [40]. For the same purpose, the newly deve-loped technique of trapping short-lived radicals by nitrones or nitroso com-pounds [40d] should be of considerable interest, as should also the chemically induced nuclear spin polarization (CINP) phenomenon [40e] be.

4.1. The Nature of the Electrochemical Process

In order to provide an introduction to the concept of *electrode potential,* it is illustrative to consider a class of reactions which is familiar to organic chemists, namely dissolving metal reductions [41]. In such a reaction, an electron from an orbital in the metal is transferred to the *lowest empty molecular orbital* of the organic substrate with formation of a radical ion. Now every organic chemist knows that a particular metal has different reducing power towards different functional groups, enabling us, *e.g.,* to reduce a carbonyl function without re-ducing a carbon-carbon double bond in the same molecule. Naively expressed, we can say that a given metal can exert a certain "electron pressure" towards a substrate, but that the electron pressure is not always high enough to allow transfer of an electron to an empty orbital of sufficiently high energy. If we want to overcome this energy difference, we will have to use a metal capable of exerting a higher electron pressure, i.e., use a more electropositive metal.

In effect, a dissolving metal reduction is an electrolytic process, except that the metal (the "electrode") is converted to a higher oxidation state and hence is consumed in the reaction. For a given metal, the electron pressure towards a given substrate is constant; in order to accomplish more and more difficult reductions, we will have to use more and more electropositive metals.

In electrochemistry, the electrode metal is not consumed during the reaction — unless we do not deliberately arrange to have it consumed, e.g., in the preparation of organometallics (Sect. 17.) — and electrons are supplied from an external DC power source. The electron pressure from the electrode (the cathode in a reduction process) can be changed at will within the limits given by the properties of the electrolyte by changing the *electrode potential*, the potential change at the interface between electrode and solution. Hence a cathode can acquire the reducing properties of any of the metals used in dissolving metal reductions by simply varying the electrode potential.

It is more difficult to find a good analogy between an electrochemical oxidation process and a chemical one. Certain metal ion oxidations, e.g. by Mn (III) [42] and Co (III) [43], are known to be of the electron transfer type, one electron being transferred from the *highest filled molecular orbital* of the organic molecule to an orbital of the metal ion. The oxidizing power of a particular metal ion towards a substrate is dependent on the energy difference between these orbitals.

At an anode, electron transfer occurs from substrate to the metal, and also here the electrode potential can be varied within a very wide interval with a concomitant variation in oxidizing power.

4.2. Single-Sweep Voltammetry

Now let us examine a typical case of electrolysis in an organic system. The circuit in Fig. 1 allows us to pass a current from an external DC power source through an electrolyte by means of the *working* and *auxiliary* electrode. As working electrode we denote the electrode which supports the process that we wish to study (in Fig. 1, the anode) and we have made provision that the potential between the working electrode and a third electrode, the *reference* electrode, to be placed in the solution as close to the working electrodes as possible, can be measured by a voltmeter of very high internal resistance. The voltmeter reading then directly gives the value of the electrode potential. The reference electrode can be a saturated calomel electrode (SCE) of the kind used in pH-instrumentation, some type of silver-silver ion electrode, or any other stable electrode.

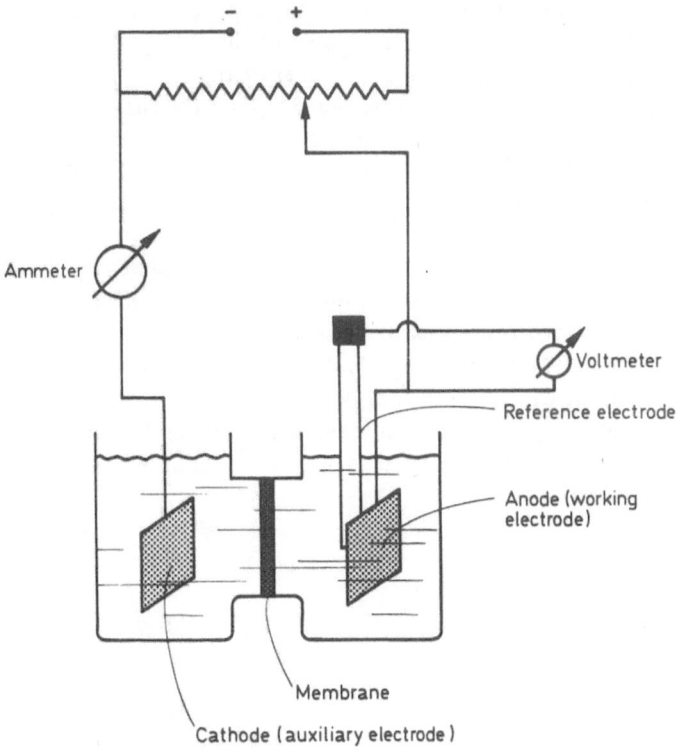

Fig. 1. Schematic drawing of electrolysis circuit and cell

To start with, the cell is filled with a solution of an electrolyte, the *supporting* or *background electrolyte,* in a suitable solvent. The role of the supporting electrolyte is primarily to make the solution conducting for the electric current, but in many cases it also serves as a reagent. The solvent may be an inert one or may take part as a reactant in the process. In the example to be discussed, we want both solvent and supporting electrolyte to be as inert as possible towards the anode; as an example of a good solvent-supporting electrolyte system (SSE) for our anodic process sodium perchlorate-acetonitrile can be mentioned; see however Ref.[465].

As shown schematically in Fig. 1, the anode and cathode *compartments* are separated from each other by a *membrane* (see Sect. 6.1), which is not permeable to organic compounds. This arrangement prevents products from either electrode process to come into contact with the other electrode. Care is taken that the anode is in contact with fresh electrolyte all the time, *e.g.,* by stirring the solution or rotating the electrode.

Now we apply a potential difference between the working and auxiliary electrode and take readings of the ammeter and the voltmeter. Note that the potential between working and auxiliary electrode, the *applied voltage,* is of no interest in this connection, since it is composed mainly of the voltage drop over the electrolyte and membrane (due to the ohmic resistance of the solution) apart from the sum of the two electrode potentials (see Fig. 3). After the first reading, the applied voltage is increased in small increments and readings taken every time. The current, or better, the *current density* (current per unit area of electrode) is then plotted *vs.* anode potential to give a *current-voltage curve* or *polarogram* of the SSE (see Fig. 2). Here we see that very little current *(residual current)* flows up to a potential of *ca.* 2.5 V, where the current starts to increase steeply. This is the *decomposition potential* or anodic limit of the SSE; at this potential an electrochemical process involving either or both components of the SSE starts to take place. The anodic limit is a rather vaguely defined concept, but is normally taken to be the potential at which the current density at a given electrode material exceeds a certain value, *e.g.* 1 mA/cm^2 in a system of preparative interest[44]. For systems of analytical interest this value is placed several orders of magnitude lower (*e.g.,* 1 μA/cm^2).

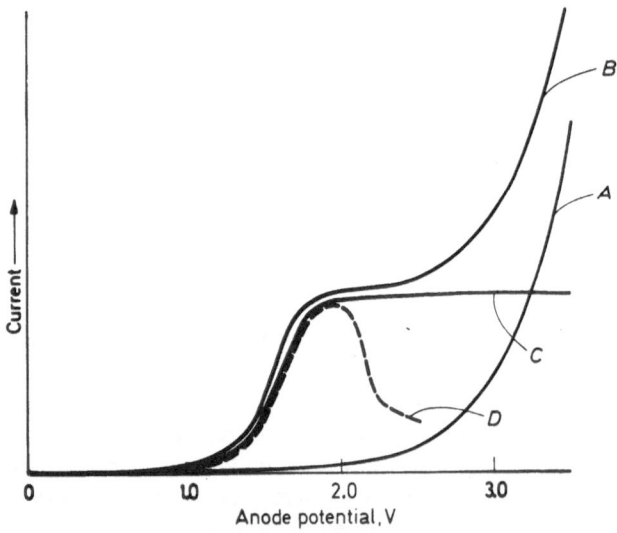

Fig. 2. Curve *A:* Polarogram of SSE; Curve *B:* Polarogram of SSE + substrate; Curve *C:* Polarogram of substrate; Curve *D:* Peak polarogram of substrate.

We now repeat the measuring procedure in the same manner, but with an organic substrate added to the SSE. The polarogram in a standard case now looks as curve *B,* Fig. 2. It is evident that a new anodic process involving the substrate takes place in the system, since the current starts to rise at a much lower anode potential. If the concentration of substrate is not too high, we observe that the current reaches a plateau value, the diffusion-controlled current. At the plateau, the rate of the electrochemical reaction is controlled by the rate of the diffusion of substrate to the anode; if all other variables are kept constant, the plateau current is proportional to the concentration of the substrate. As the potential is further increased, the current remains constant at the plateau value in a certain potential range, but finally increases steeply due to the SSE decomposition reaction.

Curve *C* of Fig. 2 is the *polarogram* of the organic substrate and is simply obtained by substraction of curve *A* from curve *B*. It is most commonly denoted the *polarographic* wave of the substrate. On a microscale it can easily be obtained using automatic equipment (a polarograph [44a]). In this case the most commonly used type of anode is the rotating platinum anode, whereas cathodic processes preferentially are studied at the dropping mercury cathode. A polarogram can also be recorded for a macro-scale electrolysis, if one employs a suitable potentiostat (see Sect. 6.1) as the DC source.

Fig. 2a shows an actual example of a polarogram run on a microscale (oxidation of anisole in acetic acid/sodium acetate) and on a macroscale (oxidation of durene in acetonitrile/sodium perchlorate). In the latter case the substrate concentration is too high for the plateau value to be reached.

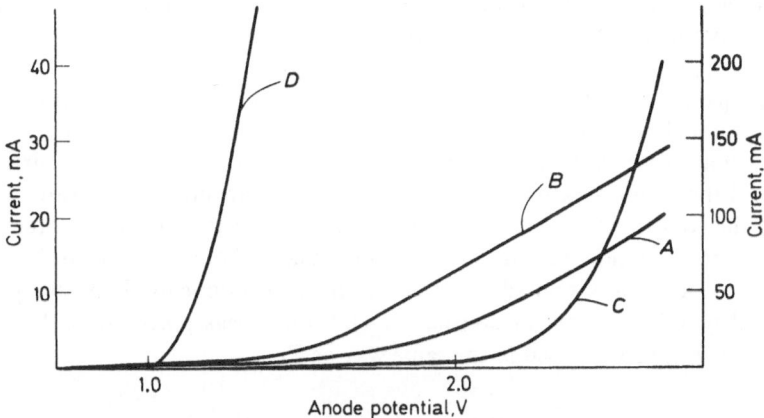

Fig. 2a. Current-voltage curves for acetic acid - 0.5 M sodium acetate *(A)*, acetic acid - 0.5 M sodium acetate - 10mM anisole *(B)*, acetonitrile - 0.5 M sodium perchlorate *(C)*, and acetonitrile - 0.5 M sodium perchlorate - 0.20 M durene *(D)*. Left ordinate axis corresponds to curves *A* and *B*, For *A* and *B*, anode potential is given *vs.* the SCE; for *C* and *D*, *vs.* Ag/0.1 M Ag$^+$

In Fig. 3, the relation between the potential differences referred to above and cell dimensions is indicated. Note that the electrode potential lies across a very short distance, of the order of 10 A, thus creating a very strong electric field in the immediate vicinity of the electrode surface (around 10^7 V/cm).

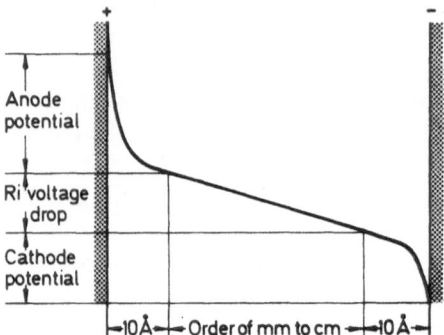

Fig. 3. Showing the relation between potential differences and cell dimensions. In a typical case, the anode potential may be 2.0 V, the cathode potential 1.5 V, and the Ri drop 20 V, giving an applied potential of 23.5 V

The polarogram of Fig. 2 (curve *B*) was recorded in a stirred solution, in which fresh electrolyte is brought into contact with the anode constantly. If one employs a *stationary* anode in an *unstirred* solution, the polarogram does not display a plateau but instead a *peak*, since now the solution near the electrode becomes depleted in substrate as the anode potential is increased (curve *D*, Fig. 2). Such a polarogram is called a *peak polarogram*.

A polarogram of the type shown in Fig. 2 immediately tells us that the substrate is the electroactive species in a certain potential region below that of the anodic limit of the SSE. The potential at half the plateau value of the current is denoted the *half-wave potential* ($E_{1/2}$) of the substrate and is a measure of how easily the compound is oxidized. With a knowledge of the half-wave potential of the substrate it is now easy to link products isolated from macroelectrolyses with the electrode processes possible in the system. This is done by the technique of controlled potential electrolysis (Sect. 4.4). From a peak polarogram, the peak potential (E_p) may be used in the same way as $E_{1/2}$.

4.3. Cyclic Voltammetry

The kind of voltammetry described in Sect. 4.2. is of the single-sweep type, *i.e.*, only one current-potential sweep is recorded, normally at a fairly low scan rate (0.1–0.5 V/min), or by taking points manually. Cyclic voltammetry is a very useful extension of the voltammetric technique. In this method, the potential is varied in a cyclic fashion, in most cases by a linear increase in electrode potential with time in either direction, followed by a reversal of the scan direction and a linear decrease of potential with time at the same scan rate (triangular wave voltammetry). The resulting current-voltage curve is recorded on an XY-recorder, in which case scan rates of up to 50 V/min can be employed. For higher scan rates, the cyclic polarogram has to be displayed on an XY-storage oscilloscope.

The usefulness of cyclic voltammetry (CV) lies in the possibility of estimating the stability of intermediates formed in an electrode process. If, for example, an intermediate is formed on the first anodic sweep and is stable towards the reagents present on the time-scale used, it will be reduced back to starting material on sweeping in the cathodic direction (see curve A, Fig. 4), which shows up as a matching cathodic peak in the cyclic polarogram. On the other hand, if one adds a reagent which reacts rapidly with the intermediate, the cathodic peak disappears completely (curve B, Fig. 4) and other peaks, due to the products formed, may appear in the cyclic polarogram. Thus one can immediately see whether an electrode process is of the reversible (curve A) or irreversible (curve B) type.

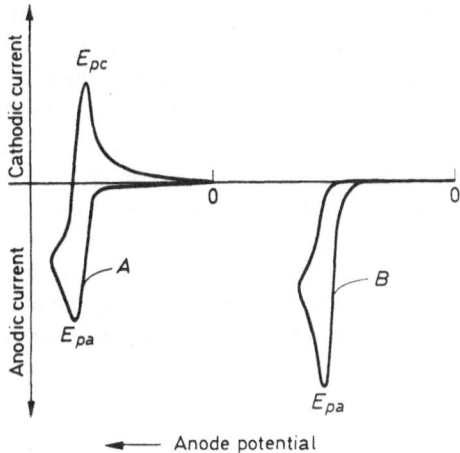

Fig. 4. Cyclic polarograms of the reversible (curve A) and irreversible type (curve B). The potential scan starts at the 0-point in the anodic direction and the scan direction is reversed a few tenths of a volt after passing the peak potential E_{pa}

Both single-sweep and cyclic voltammetry can provide information about the approximate number of electrons transferred in each wave or peak. This is done by comparing the plateau or peak height with that of a known one- or two-electron transfer process under identical conditions (as an example, the oxidation of 9,10-diphenylanthracene to the cation radical is a commonly used reference reaction).

We shall not attempt at a detailed discussion of voltammetry here. Instead the reader is referred to Adam's book [25] in which both theoretical and experimental aspects are treated thoroughly. It only remains to be stressed that voltammetry — especially cyclic — is probably the most powerful technique available for study of organic electrode processes at the level organic chemists are interested in. Cheap instrumentation is commercially available, at least as long as one is interested in attaining modest sweep rates only (up to 50 V/min).

4.4. Controlled Potential Electrolysis

Once voltammetry has provided information about the electrode processes possible in a given system, it remains to find out how products are connected with these. This is done by running a macro-scale electrolysis in which the working electrode is kept at a constant potential, this being chosen in a range where only one process takes place. Any products isolated from such an electrolysis must necessarily originate from the process under study. To see why it is so important to keep the potential constant, let us consider the polarograms in Fig. 5. These have been run at two different concentrations of the same substrate under otherwise identical conditions. Suppose for a moment that we choose to run an electrolysis employing a *constant current*, i_c, at the value of the plateau current of the upper curve of Fig. 5. After a while, part of the substrate has been consumed in the electrode reaction and a polarogram now looks as the lower curve. Since the current has been kept constant at i_c during electrolysis, the electrode potential has now changed to a value where the SSE decomposition process can take place simultaneously with the substrate process which we are interested in. This means at best that current is being wasted on the SSE process, at worst that side-reactions occur due to the mixing of two processes. Thus, it is not possible to trace products from a constant current electrolysis to a particular electrode process.

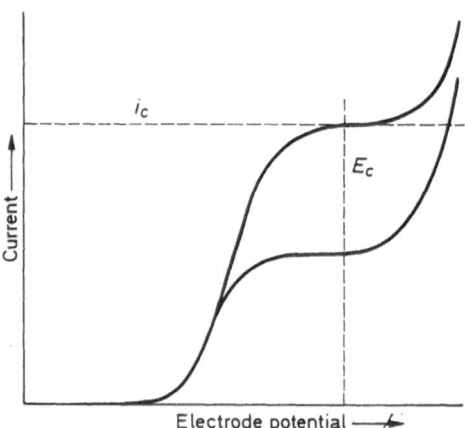

Fig. 5. Current-voltage curves recorded at two different substrate concentrations; operation at constant current or constant potential

On the other hand, if the electrolysis is performed at a constant potential, E_c, chosen in a potential region where only the substrate process takes place (see Fig. 5), the current will decrease with time to a very low value, signalling the endpoint of the reaction. No products from the second process will be formed, since this takes place in a higher potential region.

Anodic acetoxylation is an illustrative example of these principles. Anodic oxidation of sodium acetate in acetic acid at a platinum anode under constant current conditions yields ethane in almost quantitative yield. The mechanism was supposed to be discharge of acetate ion at the anode with formation of an acetoxy radical, which subsequently would undergo decarboxylation with formation of methyl radicals as shown in Eqs. (14) and (15).

$$CH_3COO^- \longrightarrow CH_3COO \cdot + e^- \qquad (14)$$

$$CH_3COO \cdot \longrightarrow CO_2 + CH_3 \cdot \qquad (15)$$

To demonstrate the intervention of acetoxy radicals, an aromatic substrate (anisole or naphthalene) was added to the electrolyte and the reaction run under constant current conditions. The isolation of aryl acetates from this reaction was considered as evidence for the intermediacy of acetoxy radicals, the aryl acetate being formed *via* a homolytic attack of the acetoxy radical on the aromatic compound [45-47].

However, under close examination this reaction was found to correspond to the simple case shown in Fig. 2 [48-50]. Sodium acetate-acetic acid has an anodic limit of about 2.0 V *vs.* SCE, whereas both anisole and naphthalene (and a large number of other substrates) have half-wave potentials far below this value. Controlled potential electrolysis (cpe) at low anode potentials showed that aryl acetates indeed were formed *via* discharge of the aromatic compound and not acetate ion. Another case in which the mechanism is clearly indicated by results from cpe is anodic acetamidation of alkylaromatic [51] and aliphatic compounds [44].

The use of the cpe technique becomes even more important in cases where the polarogram of the substrate displays two separate waves, indicating that the compound undergoes two successive electron transfers. Thus the polarogram of 4,4'-dimethoxystilbene (3) shows two anodic waves in acetonitrile-sodium perchlorate, with half-wave potentials of 0.90 and 1.15 V *vs.* SCE, respectively. Cpe of 3 in CH_3CN-CH_3COOH-NaOAc at low potential (< 1.05 V) gave exclusively a product (4) resulting from a one-electron oxidation, whereas at higher potentials (> 1.20 V) a two-electron oxidation product (5) was formed [52]. At intermediate potentials, both products were formed.

Similar results were obtained in the anodic oxidation of 9,10-dimethylanthracene [53].

It is obvious that two waves in a polarogram need not necessarily be due to a single compound, but also to a mixture of two substrates each displaying a single

21

$$MeO-\langle\ \rangle-CH=CH-\langle\ \rangle-OMe \xrightarrow{-e^{\ominus}} MeO-\langle\ \rangle-\overset{+}{CH}-\overset{\cdot}{CH}-\langle\ \rangle-OMe$$

3

1) Dimerization
2) Cyclization
3) OAc⊖

$\downarrow -e^{\ominus}$

$$R-\overset{+}{CH}-\overset{+}{CH}-R$$

$\downarrow OAc^{\ominus}$

OAc structure:

$$\begin{array}{c} OAc \\ MeO-\text{(naphthalene ring)}-R \\ R \quad R \end{array}$$

4

$$\begin{array}{cc} OAc & OAc \\ | & | \\ R-CH & -CH-R \end{array}$$

5

$$R=CH_3O-\langle\ \rangle-$$

wave. If the two substrates can be brought into reaction with each other *via* an electrode process, cpe gives the possibility to study the factors controlling the number of products, as for example in cathodic mixed hydrodimerization [54].

It now remains to treat cases where polarograms may actually be completely misleading with regard to the synthetic utility of a given process, but where nevertheless cpe might be of great help. These are the cases where the polarogram of the SSE is not changed by addition of the organic substrate, *i.e.*, it appears as if the SSE electrode process takes place at a lower (on the anodic side) or higher (on the cathodic side) potential than the substrate reaction. Thus one would be forced to conclude from voltammetric studies that the desired process cannot take place or take place to a small extent only due to competition from the SSE process.

However, this is not always the case. As an example, electrochemical cyanation [55-60] of an aromatic compound can be carried out by anodic oxidation in methanol-sodium cyanide (Eq. (16)). The current yield (the yield of cyanation product based on the amount of

$$\text{Ar-H} + \text{CN}^- \xrightarrow[\text{Pt or C}]{\text{MeOH}} \text{Ar-CN} + \text{H}^+ + 2e^- \qquad (16)$$

charge passed through the electrolyte) is as high as 50% for anisole, although the anodic limit of the SSE is around 0.8 V and the half-wave potential of anisole is around 1.6 V. Cpe shows that no cyanation occurs at potentials below 1.4 V, whereas a 50% current yield is realized above 1.6 V. Thus the cyanation product is formed via discharge of the substrate, something which voltammetric studies cannot give any information about. Similar cases are found in the anodic substitution by cyanate [61] and methoxide ion [62,63] (see also Sect. 9.1). Another notable case is the Kolbe reaction — oxidation of a carboxylate at a platinum anode — which proceeds in a very high current yield in aqueous solution at a potential above 2.2 V *vs.* SCE, although the anodic limit of aqueous systems at platinum anodes due to oxidation of water to oxygen normally is lower than 1.7 V.

A possible cause for the suppression of a low-potential process in favor of a high-potential one is considered to be preferential adsorption [21,63] of the substrate of the high-potential process on the electrode surface. The electrode then becomes blocked for the other process. The role of adsorption for the synthetic outcome of an organic electrode process is difficult to ascertain at the present state of our knowledge but might well be very important. In the following, we shall have occasion to discuss several reactions in which products of an unusual nature are found and where adsorption might be the cause of their formation.

For large-scale processes, which can be run continuously, there is no need for potential control since the substrate concentration can be kept constant all the time. This in effect serves to keep the electrode potential constant.

A recent review on cpe is available [64], as well as several older ones [6,8].

4.5. Coulometry

From Figs. 2 and 5 it is evident that cpe in a range where only one electrode reaction occurs can give a measure of the amount of charge (in Coulombs or As) necessary to convert all of the electro-active compounds to products, and hence the number of electrons consumed per molecule of electro-active material. This can be done very simply by plotting current readings (in A) *vs.* time (in s) from a cpe electrolysis and estimating the area under the curve, or by using an electronic integrator or a copper or oxyhydrogen coulometer. From a knowledge of the material balance the current yield of product(s) can be calculated.

4.6. ESR Spectroscopy

As mentioned before, ESR spectroscopy has been used extensively for the study of electrochemically generated radicals and radical ions [40]. A word of caution is necessary with regard to the interpretation of such results; the detection of a particular radical species is no definite proof that the radical is an intermediate in the formation of products. This can only be established by supporting the ESR studies by kinetic investigations. Also the failure to detect radicals from an electrode process does not mean that radicals are not intermediates, only that they may be too short-lived to be detectable. Generally, one can estimate the lower limit for detection of radicals from electrode reactions at a half-life of about 0.1 sec for external generation and 0.01 sec for internal generation.

5. Experimental Factors

The following experimental variables are of importance for the outcome of an organic electrosynthesis:

1. The nature of the substrate,
2. The nature of the SSE,
3. The electrode material,
4. The electrode potential,
5. The temperature.

We shall now examine how these factors can influence an electroorganic reaction, keeping in mind that they usually are interconnected in a not necessarily simple or predictable way.

5.1. The Substrate

The first question one asks when considering an electrochemical transformation of any kind is whether the substrate is reducible or oxidizable within the accessible potential range (as determined by the SSE), from about -3.3 V at the cathode to about +3.7 V at the anode (*vs.* SCE). This is best done by studying the electrochemical behavior of the compound using any of the simple voltammetric techniques, for a cathodic reaction at the mercury electrode, for an anodic one at a platinum anode. An inert SSE with low cathodic or high anodic limit, respectively, is chosen for this purpose (see Sect. 5.2 and Table 1), since one is only interested in finding a value of $E_{1/2}$ or E_p of the substrate at this stage of the investigation. However, the use of an inert SSE sometimes presents problems due to film formation at the electrode surface. Since there is no nucleophile or electrophile present in the SSE to react with the intermediate formed in the electron-transfer process, polymerization will occur with resultant formation of a non-conducting film on the electrode. Thus, one is sometimes forced to use a less inert SSE in order to obtain useful voltammetric results, but should then keep in mind that cyclic voltammetry might not give any information regarding otherwise relatively stable intermediates, since they will be consumed in fast reactions with any of the components of the SSE. Also, a reaction which displays two successive waves in an inert SSE might give only a single wave in a less inert SSE, as for example in the reduction of aromatic hydrocarbons [65], ketones [65a], and azo compounds [65b] in the presence of a proton source.

With the wide range of SSE:s presently available, it should be possible to get an experimental value of $E_{1/2}$ or E_p for almost any substrate, except possibly for those which are extremely difficult to reduce or oxidize or tend to form films. In the rare cases where an experimental value cannot be obtained, a reasonable value can often be inter- or extrapolated using known correlations between Hückel MO parameters and oxidation or reduction potentials, or between gas phase ionization potentials and oxidation potentials [66]. A very thorough discussion of structural effects on electrode reactions is available [24], as well as a comprehensive list of oxidation potentials of organic compounds [10].

Some rules of thumb are useful for quick estimation of how easily a substrate is oxidized or reduced, respectively. Following the postulate that anodic electron transfer occurs from the highest filled MO of the substrate, it is obvious that groups which raise the energy of this MO will lower oxidation potentials. These are the same groups that stabilize cationic centers inductively and/or by conjugation, *e.g.,* alkyl groups, unsaturated hydrocarbon groups, aryl groups, alkoxy, and hydroxy groups, amino or substituted amino groups, and halogen atoms. On the contrary, since cathodic electron transfer occurs to the lowest empty MO, substituents which lower the energy of this MO will raise reduction potentials. These are the electron-withdrawing substituents, such as the nitro, carbonyl, cyano, and sulphonyl groups, which stabilize carbanionic centers.

Having obtained an approximate measure of $E_{1/2}$ or E_n for the substrate, one can then compare it with the electrochemical properties of the SSE to be used. If the limit of the SSE is outside the range in which the substrate reacts one can be fairly sure that cpe will give products resulting from the substrate electrode process. If this is not the case, one should not be discouraged from running a preparative experiment at a series of different potentials, as has already been pointed out (Sect. 4.4).

5.2. Solvent-Supporting Electrolytes

Of all factors involved, the SSE plays one of the most decisive roles for the synthetic result of an organic electrode process. In the first place, it serves as a medium for the reaction and hence should possess good solvent properties with respect to organic compounds. In the second place, it should have good conductivity for the electric current, *i.e.,* the solvent should have a reasonably high dielectric constant and the supporting electrolyte be present in fairly high concentration (at least 0.1 M). Thirdly, it may serve as a source of reactant in the chemical processes following electron transfer. Table 1 lists a number of nonaqueous SSE:s of interest in organic electrochemistry [66a], together with their anodic and cathodic limits on platinum. It should be pointed out that the values given in Table 1 refer to carefully purified reagents and usually are defined for analytical work (*i.e.*, limits are given as the potentials at which the current exceeds a few $\mu A/cm^2$). They are rather sensitive to traces of impurities, especially water, and one should not be surprised to find somewhat differing values for an SSE prepared from good quality commercial reagents.

As can be seen by inspection of Table 1, the anodic and cathodic limit of a particular SSE depends on an electrochemical process involving either solvent or supporting electrolyte. In for example acetonitrile the anodic limit is dependent on the nature of the anion and the cathodic limit on the nature of the cation, whereas in dimethyl sulfoxide (see Table 2) the anodic limit is due to oxidation of solvent in cases where an oxidation-resistant anion is present, and the cathodic limit is dependent on a process involving reduction of the cation. Thus, one can order anions in a series of increasing resistance towards anodic oxidation:

$$I^- < Br^- < Cl^- < NO_3^- < CH_3COO^- < ClO_4^- < BF_4^- < PF_6^-$$

Table 1. *Accessible potential ranges of non-aqueous solvent-supporting electrolyte systems on platinum (in V vs. SCE)*

Solvent	Electrolyte	Cathodic limit	Anodic limit
Acetic acid [48]	NaOAc	-1.0	2.0
Acetonitrile [67]	LiClO$_4$	-3.2	2.6
Acetonitrile [67]	NaClO$_4$	-1.6	2.6
Acetonitrile [67]	Bu$_4$NClO$_4$	-2.7	2.6
Acetonitrile [44]	Et$_4$NBF$_4$		3.5
Acetonitrile [44]	Et$_4$NPF$_6$		3.6
Dimethylformamide [60]	Et$_4$NClO$_4$	-2.8	1.9
Dimethylsulfoxide [68]	Et$_4$NClO$_4$	-2.3	2.1
Dimethylacetamide [60]	Et$_4$NClO$_4$	-2.7	1.6
Hexamethylphosphoramide [69]	LiClO$_4$	-3.3	1.0
Methanol [60]	LiClO$_4$	-1.0	1.3
Methanol [63]	KOH	-1.0	0.6
Methylene chloride [70]	Bu$_4$NClO$_4$	-1.7	1.8
Nitromethane [71]	LiClO$_4$	-2.4	3.0
Propylene carbonate [72]	Et$_4$NClO$_4$	-1.9	1.7
Pyridine [60]	Et$_4$NClO$_4$	-2.2	3.3
Sulfolane [73]	Et$_4$NClO$_4$	-2.9	2.3
Tetrahydrofuran [60,74]	LiClO$_4$	-3.2	1.6
Tetrahexylammonium benzoate [75]	–	-1.2	0.3

Table 2. *Potential ranges in dimethylsulfoxide containing different supporting electrolytes (0.1 M) on platinum (in V vs. SCE; identical values were found on a vitreous carbon anode* [68])

Electrolyte	Cathodic limit	Anodic Limit
$LiClO_4$	-2.68	2.10
$KClO_4$	-2.33	2.10
$NaClO_4$	-2.08	2.10
KNO_3	-2.33	2.10
KBF_4	-2.33	2.10
$K_2S_2O_8$	-2.33	2.10
$LiCl$	-2.68	1.52
Me_4NCl	-2.4	1.52
Et_4NClO_4	-2.3	2.10
Bu_4NBr	-2.4	1.45

and cations in a series of increasing resistance towards cathodic reduction:

$$Na^+ < K^+ < R_4N^+ < Li^+$$

The data in Tables 1 and 2 are given for platinum as both anode and cathode material. For mercury, which is the most commonly used cathode metal, the cathodic limits are normally displaced to somewhat more cathodic potentials than on platinum. Mercury is seldom useful as anode material, since it is oxidized at potentials above +0.4 V (SCE) and goes into solution.

The last entry of table 1, tetrahexylammonium benzoate, is an example of the use of molten salts as electrolytes. In this particular case, the salt is liquid at room temperature, but it has been reported that tetrabutylammonium nitrate at 150° can be used for polarographic and preparative work [75a] (oxidation of polycyclic aromatic hydrocarbons). The use of molten salts as SSE:s is of great interest because of the high conductivities of such media as compared to conventional SSE:s and deserves further studies.

In aqueous or aqueous-organic SSE:s the accessible potential range is dependent on the electrochemical oxidation and reduction of water (or hydroxyl ions and protons in acid or alkaline media) with formation of oxygen and hydrogen, respectively. The potentials at which these processes take place are different for different electrode materials [9], in that the anodic limit for aqueous systems moves to more anodic potentials in the series

$$Ni < Pb < Ag < Pd < Pt < Au$$

and the cathodic limit moves to more cathodic potentials in the series

$$Pd > Au > Pt > Ni > Cu > Sn > Pb > Zn > Hg$$

Thus, to perform a reduction of a difficultly reducible substrate in a water-containing system one would normally prefer a metal like Pb, Zn, or Hg, whereas for difficult oxidations Au or Pt would be the anode material of choice. In non-aqueous SSE:s the choice of electrode material is not critical from this point of view.

An inert SSE of great utility in cathodic reductions is a concentrated solution of a hydrotropic quaternary ammoniumsalt, *e.g.* tetrabutylammonium tosylate, in water [76]. Such a solution has solvating properties approaching those of an organic solvent and hence enables one to obtain fairly concentrated homogeneous solutions of organic compounds in an aqueous medium.

In cases where the SSE does not in itself serve as a source of nucleophiles or electrophiles, such reagents can be added to the SSE, normally in large excess compared to the substrate concentration. Gaseous reagents, *e.g.,* CO or CO_2, can be bubbled over the electrode surface; however, in order to avoid waste of material, pressurized cells are preferable in such cases [77].

In general, inert SSE:s tend to favor coupling reactions between two or more substrate molecules whereas those with nucleophilic or electrophilic properties favor substitution or addition reactions. As an example the anodic oxidation of durene [78] on platinum can be controlled to give substitution product only in a strongly nucleophilic SSE (Eq. (17)) and coupling product only in a non-nucleophilic SSE (Eq. (20)). In SSE:s of intermediate nucleophilicity, both types of products are formed (Eqs. (18) and (19)).

Minor products have been excluded from the equations, since they are not connected with the phenomenon under discussion. Mechanistically, these results are best interpreted in terms of a carbonium ion intermediate, durene being oxidized at the anode to a benzylic cation. The cation then reacts preferentially with the strongest nucleophile present, which in Eq. (17) is acetate ion, in Eq. (18) acetic acid, in Eq. (19) acetonitrile, and in Eq. (20) durene itself.

However, even if a consideration of the macroscopic properties of the SSE many times is useful as a first approximation for predicting the outcome of an unknown electro-organic reaction, it must be borne in mind that the composition of the electrolyte at the electrode surface and its immediate vicinity might be completely different from that of the bulk of the solution. Current theory [19,79] assumes that the electrode surface is covered by an adsorbed layer of ions and neutral molecules during electrolysis. The thickness of this layer, the *electrical*

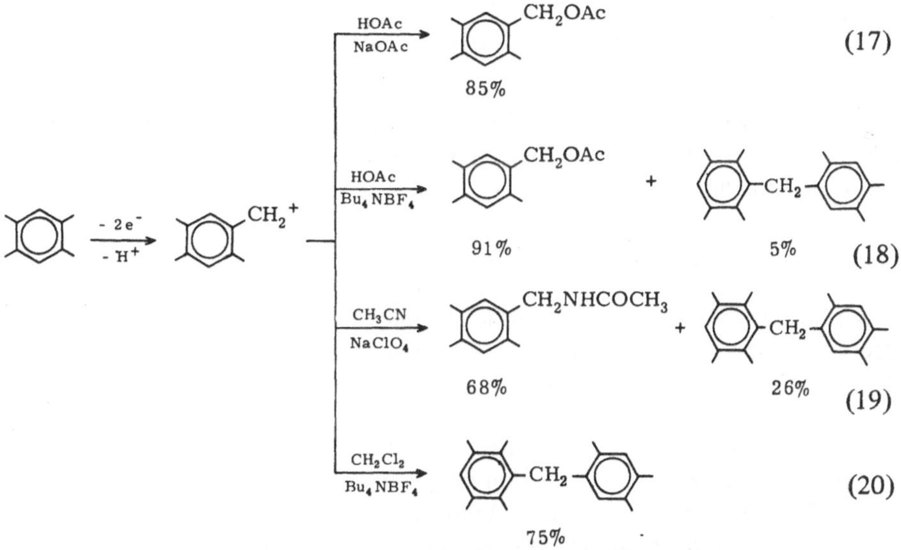

$$\text{(benzene ring)} \xrightarrow[- \text{H}^+]{-2e^-} \text{(benzene ring)}CH_2^+$$

$$\xrightarrow[\text{NaOAc}]{\text{HOAc}} \text{(ring)}CH_2OAc \qquad (17)$$
85%

$$\xrightarrow[\text{Bu}_4\text{NBF}_4]{\text{HOAc}} \text{(ring)}CH_2OAc \quad + \quad \text{(ring)}CH_2\text{(ring)} \qquad (18)$$
91% \qquad\qquad 5%

$$\xrightarrow[\text{NaClO}_4]{\text{CH}_3\text{CN}} \text{(ring)}CH_2NHCOCH_3 \quad + \quad \text{(ring)}CH_2\text{(ring)} \qquad (19)$$
68% \qquad\qquad 26%

$$\xrightarrow[\text{Bu}_4\text{NBF}_4]{\text{CH}_2\text{Cl}_2} \text{(ring)}CH_2\text{(ring)} \qquad (20)$$
75%

double-layer, is of the order of 10 A. The region between the electrical double-layer and the bulk of the solution is denoted the *diffuse layer* (50-100 A in thickness), in which concentrations gradually change from those of the double-layer to those of the bulk of the solution. Since the electron transfer process necessarily must take place with the substrate molecule situated very close to the electrode surface, any short-lived intermediate formed will undergo further chemical reactions in a medium with properties differing from those of the bulk of the solution.

A drastic example of this phenomenon is encountered in the cathodic hydro-dimerization [76,80] of acrylonitrile to adiponitrile. This can be accomplished in very high yield in a concentrated solution of a tetraalkylammonium tosylate in water. Practically no propionitrile, the product of hydrogen addition, is formed. The reaction is believed to occur *via* formation of the acrylonitrile anion radical (6), which then attacks a second molecule of acrylonitrile. Further reduction of the resulting anion radical (7) followed by protonation of the dianion gives adiponitrile (Eqs. (21), (22) and (23)).

$$CH_2=CH\text{-}CN + e \xrightarrow[\text{Pb}]{\text{Hg or}} {}^-\!CH_2\text{-}\dot{C}H\text{-}CN \qquad (21)$$
6

$${}^-\!CH_2\text{-}\dot{C}H\text{-}CN + CH_2=CH\text{-}CN \longrightarrow NC\text{-}\overset{-}{C}H\text{-}CH_2\text{-}CH_2\text{-}\dot{C}H\text{-}CN \qquad (22)$$
7

29

Experimental Factors

$$(7) \quad \xrightarrow[2 \text{ H}^+]{+\text{e}} \quad \text{NC(CH}_2)_4\text{CN} \tag{23}$$

It now remains to explain why the anion radical (6) does not react with water to form propionitrile (Eq. (24)), which in view of the high water contents of the SSE is the most likely reaction.

$$(6) \quad \xrightarrow{\text{H}^+} \quad \text{CH}_3\text{-}\overset{\cdot}{\text{C}}\text{H-CN} \quad \xrightarrow[+\text{H}^+]{+\text{e}} \quad \text{CH}_3\text{CH}_2\text{CN} \tag{24}$$

A probable assumption is that tetraalkylammonium ions are adsorbed at the electrode surface, thereby creating a region with very low water concentration in the immediate proximity of the electrode. This would favor the coupling process. In support of this hypothesis alkali metal cations, especially the strongly hydrated $\text{Li}^\oplus$, favor propionitrile formation under otherwise identical conditions[80].

Similar phenomena are observed on the anodic side, where the nature of the anion of the SSE can exert a powerful influence on the chemical follow-up reactions, as in the anodic oxidation of hexamethylbenzene in acetonitrile-water (molar ratio 9:1) [51] in the presence of different supporting electrolytes (Eq. (25), (26)). The reaction is performed at an anode potential of 1.4 V vs. SCE, far below the region where the SSE:s employed are decomposed.

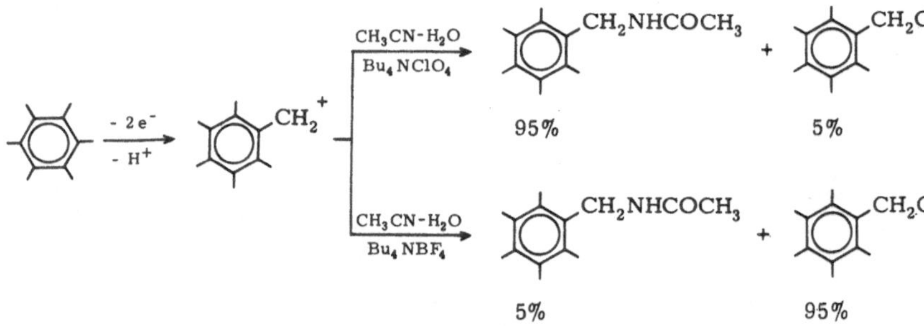

These examples clearly indicate that the composition of the SSE need not always reflect itself in product distribution, as one would predict from prevailing theory of organic reactions in homogeneous systems. There is always a possibility that some species of the SSE is available in much higher or lower concentration at the electrode-solution interface. This seems to be especially important in the case of water which can strongly influence product composition even if it is present in trace amounts only in the SSE [51]. Other hydrogen-bonding solvents show similar properties in this respect [51].

5.3. The Nature of the Electrode Material

The electrode material has always been considered to be of prime importance in determining the result of an electrochemical synthesis, as can be seen by a cursory look into the older literature [4-6]. Especially for cathodic reductions many examples are known in which the yield of the desired product is strongly dependent on the cathode material [9]. In the light of modern developments and in view of the fact that many of these electrolyses were run in aqueous or aqueous-organic systems one would guess that many of these cases actually constitute constant potential electrolyses, albeit run under constant current conditions. With a given set of experimental conditions (water, organic co-solvent, and supporting electrolyte) one has the option to change the decomposition potential of the SSE by changing electrode material (see Sect. 5.2). If the substrate is reducible or oxidizable in a potential region where the SSE decomposition process takes place a constant current electrolysis will in effect be run at an almost constant potential due to the "buffering" effect of the SSE process on the potential. Hence it is not surprising that the electrode material is of critical importance for constant current electrolyses in systems of high and relatively high water contents. It is probable that cpe using a single electrode material in an "inert" SSE of low water contents would work equally well in many of these cases.

However, also under conditions of cpe in an inert SSE some reactions are sensitive to the choice of electrode material. One such case has already been mentioned (Sect. 2; Eq. (1)); another notable case is the Kolbe reaction (Sect. 12.1a) which normally is run at a platinum anode in order to optimize the yield of product formed *via* the radical pathway (Eqs. (27) and (28)) but takes a different course on carbon anodes [82], where products derived from an over-all two-electron oxidation (the carbonium ion pathway, Eq. (29)) are formed preferentially. It has been suggested that the ability of the carbon anode to promote the generation of carbonium ions is due to the presence of paramagnetic centers at the sur-

$$RCOO^- \xrightarrow{\;-e\;} R\cdot + CO_2 \tag{27}$$

$$2\,R\cdot \longrightarrow R\text{-}R \tag{28}$$

$$R\cdot \xrightarrow{\;-e\;} R^+ \longrightarrow \text{elimination and substitution products} \tag{29}$$

face; these would impede the desorption of the initially formed radicals and therefore favor a second electron transfer [83].

Since most metals are oxidized under anodic conditions, practically useful anode materials have to be looked for among the noble metals. Platinum (also in

31

the form of platinized titanium [49]) is by far the most commonly used anode metal but also gold has found some use. Semi-conducting materials, such as graphite and lead dioxide (preferably plated onto graphite) have found wide-spread use as anodes. Boron carbide (B_4C) anodes have been used in voltammetric studies [25,84] but not explored yet in syntheses.

Cathode materials of common use in the laboratory are mercury, lead and platinum.

5.4. The Electrode Potential

The importance of controlling the working electrode potential in electrosyntheses aiming at mechanistic studies has already been stressed (Sect. 4.4). In general, potential control seems to be indispensable for mechanistic investigations but in many cases one can do without it in synthetic work, once it has been established that the desired reaction is not sensitive towards the intermediates formed in the SSE decomposition process. Thus, the exploration of a reaction for synthetic purposes would follow the pattern: 1) voltammetric studies in an inert SSE, 2) voltammetric studies in the SSE of interest (if possible), 3) cpe on a semi-micro scale at different potentials, and 4) constant current electrolyses on a large scale, if the cpe experiments have indicated that this is feasible.

5.5. The Temperature

An increase in temperature increases the rate of diffusion of electroactive material to the electrode and therefore allows for higher currents to be passed through the electrolyte. Since organic electrode processes often proceed at current densities of the order of 10 mA/cm^2, an increase in cell temperature has the distinct advantage of shortening electrolysis times (remember that electrolyses are rather time-consuming; using an electrode of an area of 100 cm^2 and a current density of 10 mA/cm^2, the time for a one-electron oxidation or reduction of one mole of substance will be about 27 h). This advantage is partially offset by the experimental difficulties of retaining volatile material in the cell at higher temperatures. In the laboratory practice, temperatures around or slightly above room temperature are therefore normally employed.

Only in rare cases does the product composition change considerably with temperature, one extreme example being the anodic oxidation of 11-bromoundecanoic acid (Eq. (30))[85].

$$Br(CH_2)_{10}COOH \xrightarrow[MeOH]{65^0} Br(CH_2)_{10}COOCH_3 + MeO(CH_2)_{10}COOCH_3$$

$$\qquad\qquad\qquad\qquad\qquad\qquad\quad 71\% \qquad\qquad\qquad 2.4\%$$

$$\downarrow{\scriptstyle 50^0 \atop MeOH}$$

$$Br(CH_2)_{20}Br$$

$$64\%$$

(30)

When planning an electrosynthesis one should however always keep in mind that competing nonelectrolytic reactions can be decelerated and thus successfully suppressed by low temperature electrolysis.

6. Experimental Procedures

6.1. Laboratory Syntheses

The design of an electrolytic cell for batchwise operation in the laboratory normally does not present any problems. Many different types of cells of more or

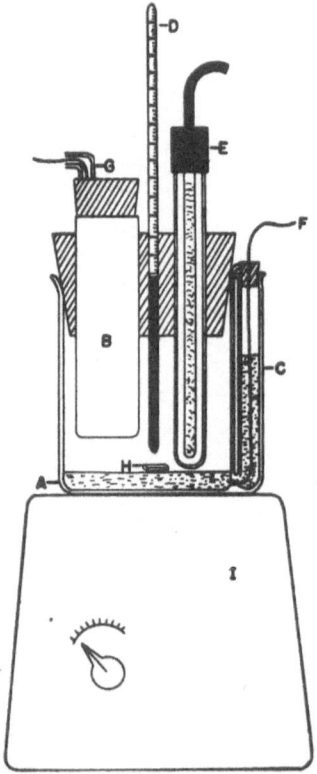

Fig. 6. Beaker type cell. *A* Pyrex beaker; *B* Alundum membrane; *C* Side-arm contact to the mercury pool consisting of a piece of glass tubing through which is sealed a piece of platinum; *D* Thermometer; *E* Reference electrode; *F* Cathode lead wire; *G* Nitrogen inlet and outlet tubes (for removal of peroxides from anode compartment); *H* Glass or teflon covered magnetic bar; *I* Magnetic stirrer. (Taken from Ref. [6], p. 34)

less elaborate construction have been described in the literature [6-8], and very little needs to be added here. Generally, one would advise the beginner in the field to use as simple cells as possible, since the feasibility of a new reaction is just as easily proven in a simple cell as in a more sophisticated one, as is of course also valid for a published electrochemical procedure. A single-compartment cell is particularly easy to assemble: a beaker, two electrodes, and some provision for cooling the cell is enough for the first trials.

If a cell divider is necessary (mostly important for cathode processes), a piece of glass tubing with a glass frit of medium porosity at one end is inserted into the beaker and serves as the auxiliary electrode compartment. For more elaborate designs (Figs. 7–9) the reader should consult the references given above and recent articles [86,87,366c].

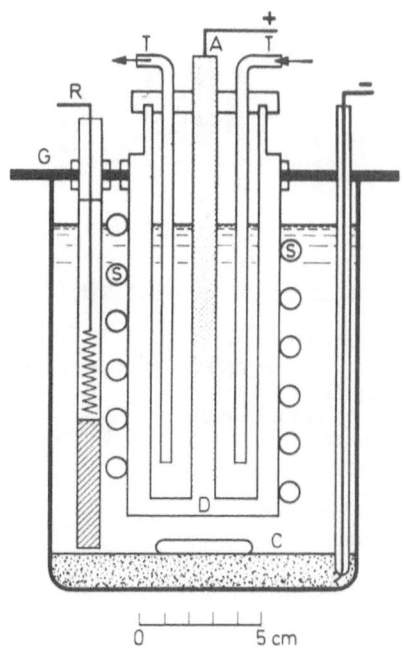

Fig. 7. Cell for macro-scale electrolysis at controlled potential consisting of a 2 l beaker covered with a glass plate G, containing holes for a silver/silver chloride reference electrode R, the anode compartment, a cooling coil S, a thermometer, an inlet for nitrogen, and one for withdrawing of samples. (The mercury cathode C has an area of 125 cm^2. The diaphragm D consists of two porous clay cylinders separated by agar containing KCl. The anolyte (15% aqueous NaOH) is continuously renewed through T. Anode A of stainless steel. The cell has been used for large scale preparations (30 to 150 g) using currents up to 25 A. From Ref. [87a]).

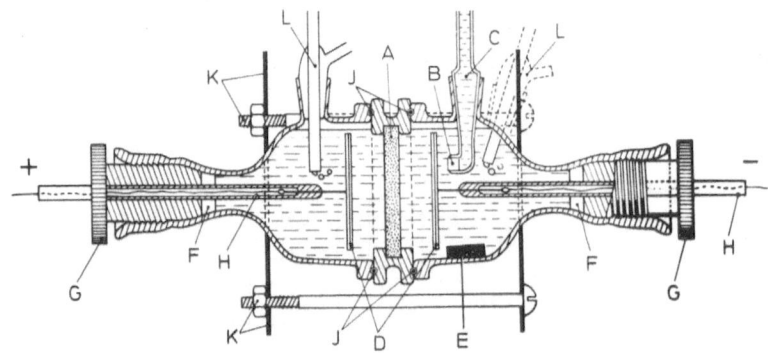

Fig. 8. Detachable diaphragm cell. (From Ref. [504])

A,B Fritted glass disk
 C   Electrolytic junction between solution and reference electrode
 D   Platinum grid
 E   Magnetic stirrer (in Teflon)
 F   Teflon gasket
 G   Torion stopper
 H   Glass tube
 J   Ground glass surface
 K   Locking screws
 L   Bubbler

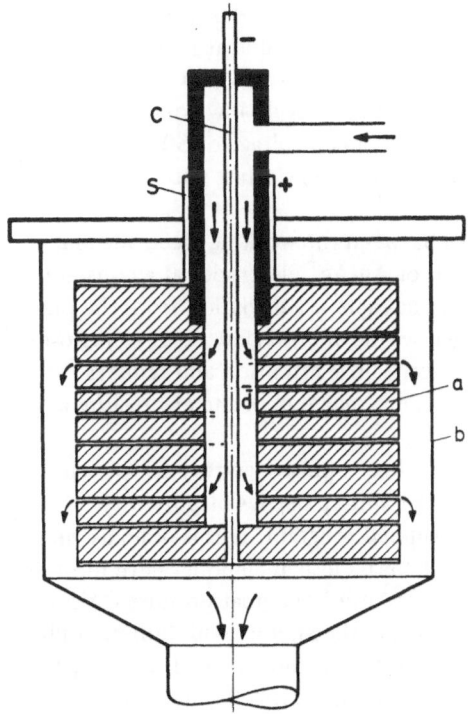

Fig. 9. Scheme of the capillary gap cell (From Ref. [366c])

a Pile of graphite plates, b Glass cell, c Current feeder, d Center of the bored plates. Cell characteristics are: high ratio of electrode surface to cell volume and small iR drop.

For continuous operation on the laboratory scale, different types of simple flow cells have been described [88,89]. In principle, this type of cell employs a bed electrode (usually in the form of a metal powder or of small glass spheres plated with a thin layer of the appropriate metal) through which the electrolyte is allowed to flow continuously. Flow cells have never achieved any wide-spread use in the laboratory, although this situation may soon change with the development of the fluidized bed electrode cell (Sect. 6.2).

New materials for cell dividers have become available commercially during recent years; these include non-fragile porous plastics membranes, cation and anion exchange membranes [89a], and ceramic diaphragms especially composed for electrolytic work.

For electrolyses without potential control any DC power source (batteries or selenium rectifiers are the simplest ones) capable of delivering 1–4 A at an applied voltage of 50–100 V will serve well for small-scale synthetic work. For electrolyses with potential control the circuit of Fig. 1 can be used, at least in principle, but of course suffers from the disadvantage of requiring continuous attention by the experimenter. Devices for automatic control of the electrode potential, *potentiostats* [89b)], are now commercially available at modest cost (ranging from $ 300 to $ 3,000) in many different sizes (ranging at modest cost 20 W to 2 kW). For operation in organic media of relatively low conductivity a potentiostat capable of delivering 2–3 A at an applied voltage of 50–100 V is sufficient for most purposes. The construction of potentiostats with fast response and high output has been described in detail [90)].

A potentiostat maintains the potential between the working electrode and a suitable reference electrode (*e.g.*, the SCE or Ag/Ag^+ electrode) at a constant preset value during the electrolysis. As soon as there is a deviation from the preset potential, the applied voltage is changed within milliseconds so as to restore the original value. The instrument requires very little attention and reduces the work of running electrosyntheses to a minimum: make-up of the electrolyte and work-up of the reaction mixture.

An interesting extension of the cpe technique is pulse electrolysis. The electrode is maintained not at one single potential, but at a series of potentials of controlled duration according to a predetermined program. This is done by means of a pulse generator (also commercially available). Pulse techniques have hitherto been used mainly for mechanistic studies [91,92)] but hold great promise for synthetic applications too [90,265)]. As an example, in the anodic oxidation of aliphatic hydrocarbons in non-aqueous medium at a platinum anode, the electrode activity falls rapidly with time if the potential is kept constant, probably because of the formation of an adsorbed film of intermediates or products. However, regular, short cathodic pulses reactivate the anode and the reaction proceeds without difficulties [30)].

6.2. Large-Scale Electrolyses

For industrial scale (and large-scale laboratory) electrolysis, batch-wise operation with potential control becomes difficult due to the lack of potentiostats of sufficiently high output (at present the limit of commercial potentiostats is 100 A at 20 V). Instead one prefers continuous steady state operation under carefully controlled conditions of concentration and temperature, which in effect is a cpe technique (Sect. 4.4) [93)]. Cell design is critically important in large scale electrolysis and has been thoroughly discussed be several authors [94-96,366c)]. The features of being capable of continous operation and having a large electrode area per unit volume of reactor are combined in the fluidized-bed cell [95,97,98)] in which a bed electrode is kept fluidized by electrolyte flow (Fig. 10). This cell type is currently under development and evaluation by Constructors John Brown [98)].

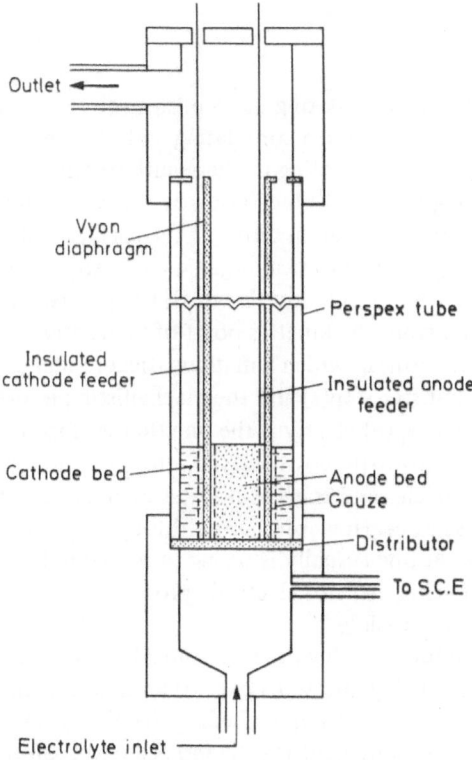

Fig. 10. Fluidized bed cell. (From Ref. [98a])

7.   General Mechanistic Considerations

The general problem of formulating mechanisms for organic electrode processes was discussed some years ago in a stimulating and thought-provoking article by Elving and Pullman [99] . According to these authors three approaches to the problem could be distinguished: a) Elucidation of the *chemical* reaction mechanism, in which the mechanism is given in terms of the species and transitions states involved using the concepts of physical organic chemistry,   b) elucidation of the *electrochemical* reaction machanism, in which the process is described in physico--chemical terms, *i.e.*, from the kinetics point of view, and   c) elucidation of the *energetic* reaction mechanism, in which a mathematical model of the relation between the electron transfer step(s) and the mechanistic picture is the ultimate goal.

Of these approaches, (a) clearly is the one that is closest to the heart of the organic chemist, used to write down curved arrow mechanisms involving the minimum number of species and transition states compatible with experimental facts at a moment's notice. However, before this approach is attempted, it is appropriate to analyze what one actually is trying to do in applying concepts from homogeneous solution chemistry to electrode processes. The overridingly important question is: Is it at all possible ?

For a complete understanding of the chemical mechanism of an organic electrode process the following questions have to be satisfactorily answered:

(a)   Is there any chemical step preceding the electron transfer step ?

(b)   Is there any requirement for the molecule to acquire a particular orientation with respect to the electrode surface for electron transfer to occur, and, if so, which is this preferred orientation ?

(c)   What is the nature of the first intermediate formed?

(d)   What kind of chemical and/or electrochemical reaction sequences does the first intermediate undergo and where do these take part with respect to the electrode?

7.1.   Chemical Step Preceding Electron Transfer

In many cases the substrate, as added from the reagent bottle, is not the electroactive species in itself, but is transformed into one by reactions taking place upon dissolution in the electrolyte, *e.g.* acid-base reactions, hydration-dehydration reactions, or cyclization reactions. The nature of the electroactive species in such cases can be elucidated by voltammetric techniques, as discussed by Zuman [24].

7.2. Orientation Effects

The question as to whether the substrate molecule has to acquire a particular orientation with respect to the electrode surface before electron transfer can occur is intimately connected with the phenomenon of *adsorption* in electrode processes. If adsorption is a necessary prerequisite for electron transfer, one would intuitively expect that this would lead to a substrate-electrode complex of defined structure (*e.g.*, an aromatic ring system would be held with its planar surface facing the electrode surface). The exact chemical consequences of this arrangement are hard to predict, but at least one implication would be one of stereochemistry. If the configuration adsorbant-surface is held for a definite period of time after electron transfer has taken place, any following fast product-forming steps would occur with the intermediate shielded from chemical attack on one side by the electrode, and therefore the stereochemistry of the reaction should be affected.

One immediately realizes that the reasoning above has at least one serious shortcoming, even if it would be correct in principle. In order to determine whether a particular electrode reaction leads to products of anomalous stereochemistry, a closely analogous homogeneous reaction must be available for comparative studies. As already pointed out, electrochemical processes often have features which are not easily simulated in homogeneous solution chemistry and hence the number of processes suitable for stereochemical studies of this kind is rather limited.

The approach outlined above has been applied in some anodic addition reactions, although the results are far from being unambiguously interpretable. As an example, Bonner and Mango [100] established that anodic oxidation of *trans*-stilbene in anhydrous acetic acid-sodium acetate gave mainly *meso*-1, 2-di-*O*-acetyl-1,2-diphenylethylene glycol (*8*) and a small proportion of *threo*-monoacetate *9*. When water was present in the electrolyte *9* was formed predominantly, together with some *dl* - *8* (presumably formed by acetylation of *9*). This result was explained in terms of an acetoxonium ion mechanism, in which the acetoxy group is added across the double bond of an adsorbed stilbene molecule to form *10*, followed by the reactions characteristic of acetoxonium ions, attack by acetate ion with inversion and water with retention:

41

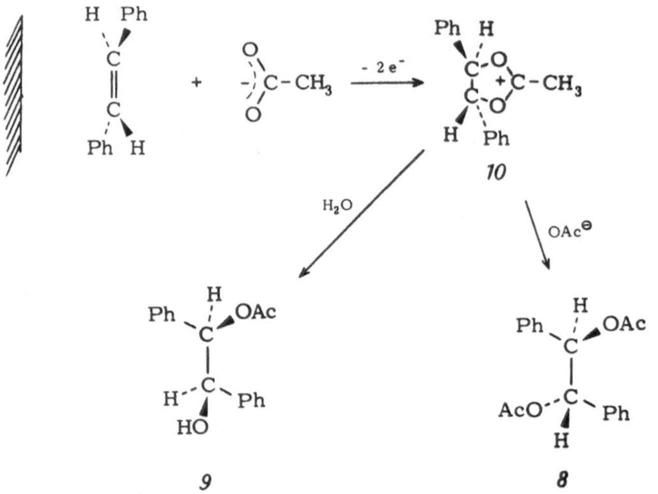

(31)

9   *8*

The stereochemistry of acetoxy group addition to *trans*-4, 4'-dimethoxystilbene [52] is analogous. However, the recent finding [101] that the analogous reaction, anodic addition of two benzoyloxy groups across a double bond, produces the same mixture of products from both *cis*- and *trans*-stilbene (stereochemistry the same as in equation (31)), would seem to make the adsorption requirement unnecessary. The stereoselectivity is then explained on the basis of formation of the thermodynamically most stable acetoxonium ion in a stepwise oxidation mechanism [101].

Anodic addition of two acetoxy groups to cyclooctatetraene in acetic acid-sodium acetate under cpe conditions proceeds in a non-stereospecific manner (Eq. (32)) in contrast to lead tetraacetate or mercuric acetate oxidation, which give exclusively *trans* addition [102].

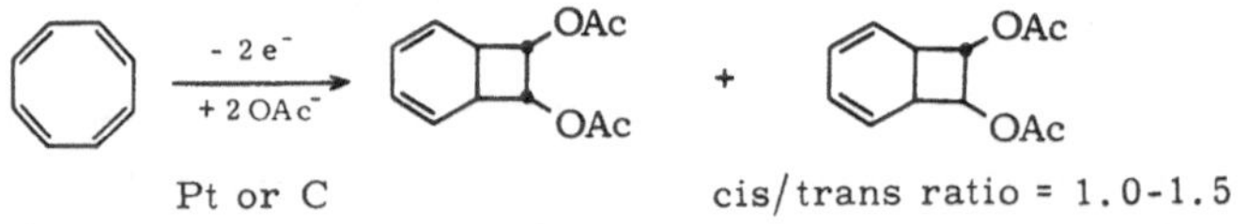

$$\text{(32)}$$

Pt or C   cis/trans ratio = 1.0-1.5

Because of the inherent complexity of the cyclooctatetraene system, these results can be explained without invoking adsorbed intermediates [102] but the possibility remains that adsorption plays an important role. By analogy with electrophilic additions to cyclooctatetraene [103], an adsorbed cyclooctatetraene molecule would be attacked preferentially from the fold of the tub conformation with formation of *endo*-8-acetoxyhomotropylium ion (*11*). Attack by acetate ion on *11* would give the *cis*-isomer, whereas the *trans*-isomer might be formed from *exo*-8-acetoxyhomotropylium ion *12* (Eq. (33)).

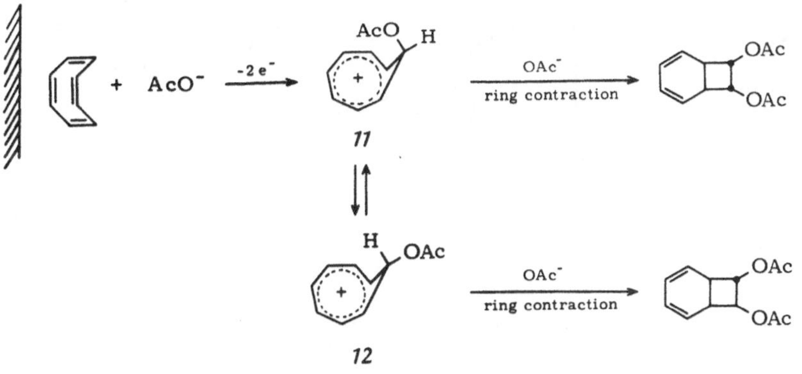

$$(33)$$

Anodic addition of two methoxy groups across the double bond of *cis*- and *trans*-stilbene has been shown to occur with a slight preference for *cis*-addition [58, 104]; again adsorbed intermediates were suggested to play an important role (Eq. (34)).

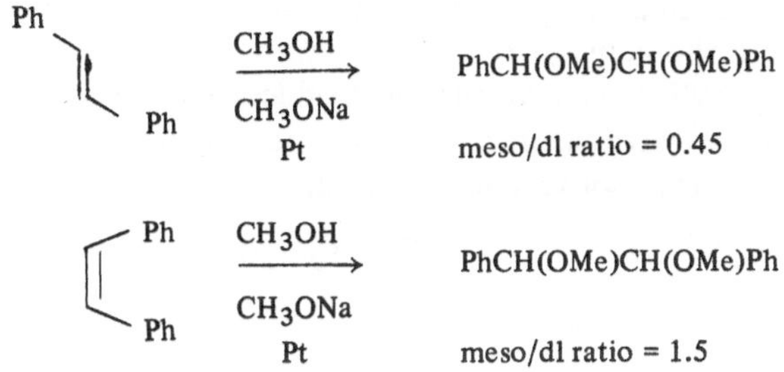

$$(34)$$

In contrast, 1,4-addition of two methoxy groups to 2,5-dimethylfuran [105] gave exactly the same mixture of isomers as chemical oxidation:

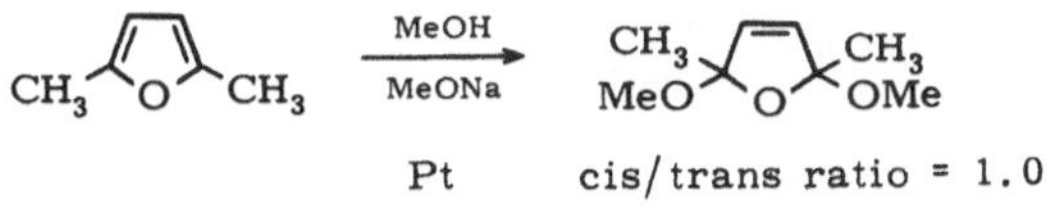

$$(35)$$

Also anodic substitution reactions can be used for studies of the possible consequences of adsorption. Thus, anodic methoxylation of N,N-dimethylbenzylamine [62,63] gives predominant substitution in the methyl group instead of the methylene group which is the site of attack in chemical oxidations. This was rationalized on the basis of adsorbed intermediates, the methyl group of the adsorbed species being more accessible for chemical attack:

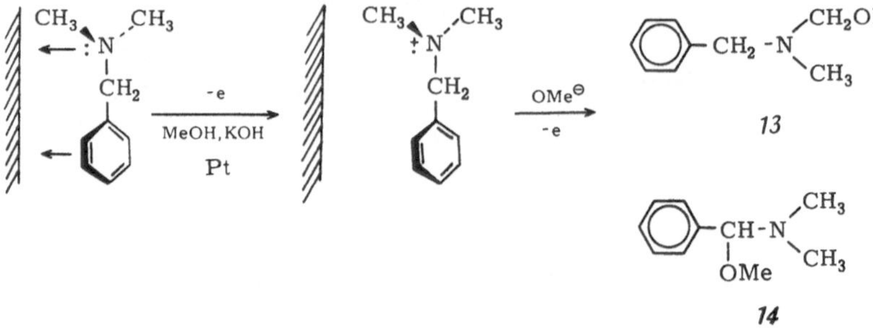

$$13/14 \quad \text{ratio} = 4$$

Another approach to the problem involves the study of molecules in which one side is more accessible for adsorption than the other one. This feature is to a certain extent present in 2-t-butylindane which would be predicted to form substrate-electrode complex *15* preferentially. On anodic acetoxylation of 2-t-butylindane a mixture of *cis*- and *trans*-1-acetoxy-2-t-butylindane in the proportions 16:84 is formed [106], in all probability *via* a carbonium ion mechanism (Eq. (37)). Generation of carbonium ion *16* by solvolysis of either *cis*- or *trans*-1-*p*-nitrobenzoyloxy-2-t-butylindane in acetic acid produces exclusively the *trans* isomer, showing that the electrochemical process indeed has different stereochemistry, although the effect is not very pronounced.

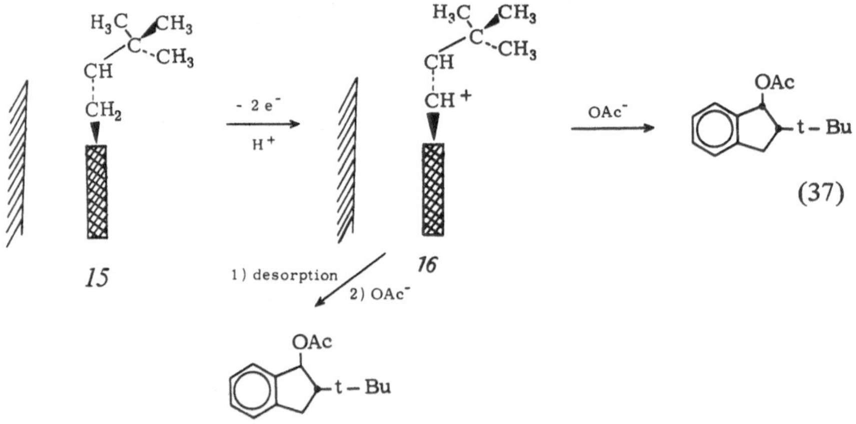

Also in some cathodic processes have orientation effects by the electrode surface been invoked, as for example in the cathodic reduction of alkynes and alkenes [107], the addition of two carboxylate groups across the double bond of *cis*- and *trans*-stilbene [108], and the cathodic reduction of geminal dihalocyclopropanes [108a, 108b].

Reduction of 4-methylcoumarin in the presence of asymmetric bases such as the alkaloids narcotine, codeine, brucine or sparteine yielded optically active 3,4-di-hydro-4-methylcoumarin in up to 19% optical yield [108d] . Similarly, the reduction of acetophenone in the presence of optically active supporting electrolytes such as ephedrin-hydrochloride produces optically active methylphenylcarbinol in 2 to 10% optical yield [108e].

Concluding, the above-mentioned results by no means constitute definite evidence that the electrode surface can exert stereochemical control of product formation. They do point in this direction, though, but much more experimentation is needed before a reasonably complete picture can emerge [108c] .

7.3. The Nature of the First Intermediate

It is generally assumed that electrons are transferred one by one (cases reported to be direct two-electron transfers [109, 110] may actually be two very closely spaced one-electron transfers [111]). This postulate immediately tells us that the first intermediate formed from a neutral substrate must be a cation radical in anodic oxidation and an anion radical in cathodic reduction and a neutral radical from oxidation of an organic anion and reduction of an organic cation. This has been amply verified both by electrochemical techniques and by ESR studies in inert SSE:s [29,65]; unfortunately, such results do not allow us to draw conclusions regarding the future reactions of these types of intermediates, as will be outlined in the following section.

7.4. Reactions of Intermediates Formed in Electrode Processes

It is natural that the organic chemist as a *first approximation* tries to interpret electrode reactions in terms of concepts derived from the study of homogeneous processes. Studies in homogeneous systems have provided an enormous amount of knowledge about the common intermediates of organic chemistry, *i.e.*, carbonium ions, neutral radicals, carbanions, and carbenes, and much of this can be directly applied to electrochemically generated intermediates of these types. However, the homogeneous solution chemistry of two very important classes of intermediates, radical anions and cations, have received much less attention [3] , and this seriously hampers our efforts to rationalize results within organic electrochemistry [111a] .

As an analogy of an electrochemical process at the cathode, consider a solution of an anion radical ($M^{\overline{\cdot}}$) salt in an inert solvent in the presence of an excess of the substrate (M) from which the anion radical is derived. The following primary reactions are then possible:

$$2\,M^{\overline{\cdot}} \longrightarrow \overline{}MM^{-} \tag{38}$$

$$M^{\cdot-} + M \longrightarrow \cdot MM^- \qquad (39)$$

$$\cdot MM^- + M^{\cdot-} \longrightarrow M + {}^-MM^- \qquad (40)$$

Only the dimerization process is considered, even if it is easily seen that the scheme may be expanded to include formation of trimers, tetramers, etc., and polymers. Thus, even in a system as simple as this the formation of $^-MM^-$ can occur *via* two pathways due to the intervention of the electron transfer reaction (Eq. (40)). This reaction scheme has been demonstrated for the radical anion from 1,1-diphenylethylene [112] (see 14.4).

A further complication is the possibility of disproportionation of the anion radical *via* electron transfer:

$$2\,M^{\cdot-} \; \rightleftarrows \; M^{2-} + M \qquad (41)$$

introducing another species, the dianion, as a potentially kinetically active species. The situation becomes even more complex when an electrophile (E^+) is present, since now the following reactions (at least) with E^+ must be considered (apart from those between E^+ and a possibly formed dimeric species):

$$M^{\cdot-} + E^+ \longrightarrow ME\cdot \qquad (42)$$

$$ME\cdot + M^{\cdot-} \longrightarrow ME^- + M \qquad (43)$$

$$ME^- + E^+ \longrightarrow EME \qquad (44)$$

$$M^{2-} + E^+ \longrightarrow ME^- \qquad (45)$$

$$2\,ME^- \; \rightleftarrows \; EME + M^{2-} \qquad (46)$$

Obviously the formation of the product, EME, can occur *via* four pathways: sequence (42), (43), and (44), sequence (42), (43), and (46), sequence (41), (45), and (44), or sequence (41), (45), and (46).

The protonation of anion radicals and dianions derived from aromatic hydrocarbons has been studied in some detail by Hoijtink and co-workers [113-115]. It was shown that apart from the reactions given above (Eqs. (42)–(46)) other disproportionation equilibria also play an important role. These are different for different anion radicals, making the whole picture very complex. Kinetic studies on the disproportionation of the nitrobenzene anion radical and some of its derivatives [116,117] have shown that in aqueous solution at a pH > 11.5, reaction (41) is of great importance, whereas the protonated radical ion and the radical ion are the kinetically active species in the pH interval between 3.2 and 11.5.

The discussion above has some interesting repercussions for cathodic processes. The reaction sequence depicted in Eqs. (47) - (49) will

$$M + e \longrightarrow M\cdot^- \tag{47}$$

$$M\cdot^- + E \longrightarrow ME\cdot \tag{48}$$

$$ME\cdot + e \longrightarrow ME^- \xrightarrow{\ E^+\ } EME \tag{49}$$

be an ECEC process (Sect. 2.), whereas a scheme involving the dianion will be an ECCC process:

$$M + e \longrightarrow M\cdot^- \tag{50}$$

$$2\,M\cdot^- \longrightarrow M + M^{2-} \tag{51}$$

$$M^{2-} + E^+ \longrightarrow ME^- \xrightarrow{\ E^+\ } EME \tag{52}$$

The distinction between these processes is not easily achieved by electrochemical techniques; only kinetic studies in homogeneous medium will tell us whether the radical anion or the dianion is kinetically active (or possibly both).

Reactions (38) - (46) have their exact counterparts in the interaction between cation radicals and nucleophiles. In the reaction between 9,10-diphenylanthracene and water [118] in acetonitrile, kinetic evidence was obtained for a rate-determining attack of water upon the cation radical, followed by reaction of the neutral radical formed with a second cation radical:

$$M\cdot^+ + Nu^- \longrightarrow MNu \tag{53}$$

$$MNu\cdot + M\cdot^+ \longrightarrow MNu^+ + M \tag{54}$$

$$MNu^+ + Nu^- \longrightarrow NuMNu \tag{55}$$

This would correspond to an ECEC process at the anode [37]. In contrast, the reaction between thianthrene cation radical and water [119] has been shown to occur with the dication as the kinetically active species due to equilibrium (56), corresponding to an electrochemical ECCC process:

$$2\,M\cdot^+ \rightleftharpoons M + M^{2+} \tag{56}$$

These examples show that our knowledge of ion radical chemistry in homogenous soluction is far from complete and that extrapolation of this knowledge to ion radicals produced at electrodes is a risky procedure, especially if one contemplates the additional complexities involved. The composition of the medium in the vicinity of the electrode is not the same as in the bulk of the solution (Sect. 5.2), the structure of the double-layer can at its best be the subject of educated guesses, and due allowance must be made for the possibility that reactions may take place between adsorbed intermediates.

8. Conversions of a Functional Group into another Functional Group

8.1. Anodic Conversions

Generally, the anodic conversion of a functional group into another one is a reaction type which has received relatively little attention during recent years. Numerous experiments have been described in the older literature [4,6,7] but the results have seldom been of any preparative value because of the lack of product selectivity. It is probable that a knowledge of half-wave potentials of substrate and product(s) in combination with cpe might improve the selectivity to a large extent in many of these cases.

Lund [120] was first in applying cpe in the oxidation of a primary alcohol to an aldehyde (which under constant current conditions would be partly or completely oxidized to the corresponding carboxylic acid) [121]. Anisyl alcohol displays two anodic waves in acetonitrile-sodium perchlorate with $E_{1/2}$ of 1.22 and 1.64 V vs. Ag/0.1 M Ag$^+$. Cpe at the plateau of the first wave (1.35 V) in the same medium consumed only 5 % of the theoretically calculated amount of electricity and no carbonyl compound was formed. Addition of a three-fold excess of pyridine (to act as a proton acceptor) gave a 72 % of anisaldehyde:

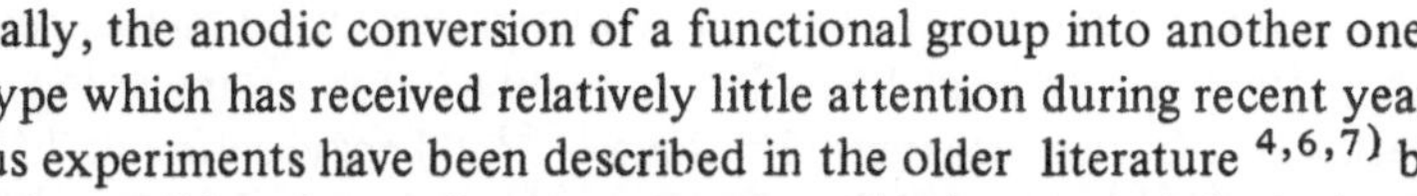

$$CH_3O-\langle\ \rangle-CH_2OH \quad \xrightarrow[\text{pyridine}]{\text{Pt, 1.35 V}} \quad 2\,H^+ + 2e + CH_3O-\langle\ \rangle-CHO \qquad (57)$$

Cpe oxidation of other alcohols (fluorenol and 2-naphthalenemethanol) were not successful due to severe filming at the electrode.

Another case of product control through cpe is the oxidation of glycerol at different potentials in H_2O-KOH [122]:

$$
\begin{array}{ccccccc}
CH_2OH & & CH_2OH & & COOH & & COOH \\
| & 0.15\text{ V} & | & 0.23\text{ V} & | & 0.39\text{ V} & | \\
CHOH & \xrightarrow{\quad\text{Ni}\quad} & CO & \xrightarrow{\quad\text{Ni}\quad} & CO & \xrightarrow{\quad\text{Pd-C}\quad} & CO \\
| & & | & & | & & | \\
CH_2OH & & CH_2OH & & CH_2OH & & COOH
\end{array} \qquad (58)
$$

Potentials given *vs.* H_2/H^+ electrode

49

The overwhelming majority of alcohol oxidations (including those of carbohydrates) have been run in SSE:s of relatively high water contents, often with stror.g acid present, under constant current conditions [123]. Selective oxidation of an alcohol to an aldehyde cannot be accomplished under such conditions; instead the carboxylic acid and its degradation product is formed. The cpe approach in SSE:s of low water contents should no doubt pay rich devidends in this area. The same applies to the oxidation of secondary alcohols, in which the acid SSE:s previously used seem to promote anodic degradation of the ketone formed.

Sulfur functions in low oxidation states have been oxidized to sulfoxides, sulfones, and sulfonic acids, often in very good yields in spite of the fact that cpe was not employed. This is probably due to the resistance towards oxidation of the products, making control of the anode potential a less critical factor, and to the use of a "potential-buffering" SSE (Sect. 5.3). Illustrative examples include the preparation of 2,2'-bishydroxyethyl sulfone [124], dibenzyl sulfoxide [125], ethanesulfonic acid [126], dibenzyl disulfoxide [125] and dimethyl sulfone [127]:

$$(HOCH_2CH_2)_2S \xrightarrow[H_2O, \, NaCl]{C} (HOCH_2CH_2)SO_2 \qquad (59)$$
$$90\%$$

$$(PhCH_2)_2S \xrightarrow[HOAc, H_2O, HCl]{Pt, \, 25^\circ} (PhCH_2)_2SO \qquad (60)$$
$$93\%$$

$$(C_2H_5S)_2 \xrightarrow[H_2O, \, EtSO_3H]{Pt} C_2H_5SO_3H \qquad (61)$$
$$80\%$$

$$(PhCH_2S)_2 \xrightarrow[HOAc, \, HCl]{Pt} (PhCH_2SO)_2 \qquad (62)$$
$$92\%$$

$$(CH_3)_2SO \xrightarrow[H_2O, \, H_2SO_4]{PbO_2} (CH_3)_2SO_2 \qquad (63)$$

A few conversions of iodo groups to iodoso and iodoxy functions have been reported [128].

Phenylhydrazine can be oxidized to phenyldiimide at a controlled potential (0.40 V vs. Ag/0.01 M Ag$^+$) in acetonitrile-lithium perchlorate under an argon blanket [129]:

$$3 \, C_6H_5NHNH_2 \xrightarrow{-2e^-} C_6H_5N=NH + 2 \, C_6H_5NHNH_3^+ \qquad (64)$$

8.2. Cathodic Conversions

The cathodic conversion of a functional group into another one is a very useful synthetic procedure, applicable to a great variety of functional groups. Also here the majority of syntheses performed prior to 1955 have been run under constant current conditions, thus making the composition of the SSE (especially with regard to pH) and the electrode material the decisive factors in the control of product selectivity. Excellent discussions of these factors and their influence can be found elsewhere [6,7,9] and need not be repeated here.

During recent years voltammetric and cpe techniques have been increasingly brought into use for the design of cathodic functional group conversions with a concomitant improvement in product selectivity. This has been especially fruitful in the rational planning of synthetic procedures for certain heterocyclic systems (see below). We shall now examine the more important developments in this field.

The polarographic behavior of aromatic carbonyl compounds at the dropping mercury electrode in aqueous SSE:s is rather complex [130]. In acid medium two one-electron waves are observed, corresponding to reduction of the protonated ketone (protonation makes the reduction easier, shifting the first wave towards less cathodic potentials with decreasing pH) and the protonated anion radical, respectively:

$$R_2C\overset{+}{=}OH \xrightarrow{\ +e^-\ } R_2\overset{\cdot}{C}\text{-}OH \tag{65}$$

$$R_2\overset{\cdot}{C}\text{-}OH \xrightarrow{\ +e^-\ } R_2\overset{-}{C}\text{-}OH \tag{66}$$

Cpe at the plateau of the first wave was reported to give the coupling product (the pinacol). At higher pH values (2-9) the pH-dependent first wave moves towards more cathodic potentials, finally merging with the pH-independent second wave. Cpe at the plateau of the combined two-electron wave gives the alcohol as product according to the mechanism outlined in Eqs. (67), (68) and (69). The protonated anion radical is now easier to reduce than the ketone, resulting in the two-electron transfer observed.

$$R_2C\text{=}O \xrightarrow{\ e^-\ } R_2\overset{\cdot}{C}\text{-}\overset{-}{O} \tag{67}$$

$$R_2\overset{\cdot}{C}\text{-}\overset{-}{O} \xrightarrow{\ H^+\ } R_2\overset{\cdot}{C}\text{-}OH \tag{68}$$

$$R_2\overset{\cdot}{C}\text{-}OH \xrightarrow{\ e^-\ } R_2\overset{-}{C}\text{-}OH \tag{69}$$

Finally, in alkaline media two practically pH-independent one-electron waves are observed, corresponding to formation of the anion radical and dianion, respectively (Eqs. (67) and (70)). The same mechanism has been proposed for ketone

$$R_2\overset{..}{\underset{\cdot}{C}}\text{-}\overset{..}{O} \xrightarrow{\;e^-\;} R_2\overset{..}{\underset{\cdot}{C}}\text{-}\overset{..}{\underset{\cdot}{O}} \tag{70}$$

reduction in non-aqueous medium [65a,131]. Addition of a proton donor (*e.g.*, phenol) displaces the second wave towards more positive potentials [65a], resulting ultimately in a two-electron wave, corresponding to the ECE process (67), (68) and (69).

Thus, the formation of an alcohol from a cathodic reduction of a ketone is favored in media where a single two-electron wave is observed, *i.e.*, in the medium pH range in SSE:s of high water contents, and in the presence of small amounts of proton donors in non-aqueous SSE:s. Alternatively, in cases where two one-electron waves are observed, the reaction should be performed at a potential corresponding to the plateau of the second wave if the cathodic limit of the SSE allows this to be reached in preparative runs (which may be difficult in aqueous SSE:s).

Relatively few investigations on large-scale cpe of ketones have been reported and the results are not wholly unambiguous. For 2-acetylpyridine it was observed that reduction at a mercury cathode in the medium pH range (3-8.5) at potentials between -0.51 and -0.99 V produced the carbinol in yields of 50-90 % with small amounts of the pinacol as a side-product [132]. In more alkaline medium (80 % ethanol-KOAc) the pinacol was the only product when the reduction was carried out at -1.2 V, whereas at -1.5 V carbinol formation became noticable. However, the same authors reported that the pinacol was the only product formed from acetophenone over the whole pH range except in two cases where it was possible to operate at slightly more cathodic potentials. It seems probable that the SSE:s employed in this investigation did not allow operation at potentials sufficiently cathodic to produce the alcohol.

Polarographic studies on the reduction of 1,3-diphenyl-1,3-propanedione in 50 % ethanol-water at the mercury cathode in the pH range 4.2 - 13.6 show that this compound is reduced in three successive steps, two one-electron transfers and a two-electron transfer at the most negative potential [133]. Cpe at either of the two first waves produces pinacolic products, whereas cpe at the third wave produced a good yield of the diol as a *meso-dl* mixture (Eq. (71)).

$$\underset{\underset{-2.0 \text{ V (SCE)}}{}}{C_6H_5\overset{\overset{O}{\|}}{C}\text{-}CH_2\text{-}\overset{\overset{O}{\|}}{C}\text{-}C_6H_5} \xrightarrow[4e^-]{4H^+} C_6H_5CH(OH)CH_2CH(OH)Ph \tag{71}$$

Cathodic reduction of nonconjugated steroidal ketones has been found to give the equatorial alcohols with a high degree of stereoselectivity and in very good yields [134]. These reactions were run at -2.6 V in aqueous ethanol-tetrabutyl-ammonium bromide. α-Methyldesoxybenzoin gave only the *erythro* form of 1,2-diphenylpropanol-1 on reduction at mercury in 40 % ethanol at pH 8 (veronal buffer) at -1.85 V (SCE) [135].

Carbonyl functions can also be reduced to methylene groups, although the optimal conditions for this reaction are not very well known [9]. The most recent investigation reported excellent yields for reduction of steroid carbonyl groups to methylene groups at mercury in an SSE made up from dioxane, water, and sulfuric acid under constant current conditions [136]. The use of deuterium oxide instead of water gave the deuteriomethylene analogs.

Carboxylic acids can be reduced in acid solution to alcohols, aldehydes, or hydrocarbons [7,9]. Polarographic and cpe studies on the cathodic reduction of isonicotinic acid in weakly acid solution (pH about 3. 1 M aqueous potassium chloride) have shown that the product, 4-pyridinealdehyde, exists as a hydrate in aqueous medium and that the hydrate does not undergo further cathodic reduction [137]. The same applies to reduction of picolinic acid [137], imidazole-2-carboxylic acid [138], and 2-thiazolecarboxylic acid [139]. In the last case the yield of aldehyde was fairly low, probably due to competing reduction of the thiazole ring. It was concluded [139] that the following requirements must be fulfilled for the facile reduction of a carboxylic acid to the aldehyde stage to take place:
a) the carboxyl group must be activated by an electron-withdrawing group
b) the aldehyde group must be protected against further reduction by forming a non-reducible derivative, *e.g,* the hydrate in aqueous solution, and c) the reduction must take place in the carboxyl group and not in other parts of the molecule.

Reduction of carboxylic acids to alcohols takes place when the intermediate aldehyde is further reducible. This has been applied to a number of aromatic acids [140,140a].

The cathodic reduction of aromatic carboxamides and imides in acid solution at lead or mercury cathodes generally leads to amines and isoindolines, respectively [7,9]. These reactions are of great preparative interest but their mechanism have not been examined. Cpe of isonicotinic amide and 2-thiazolecarboxaldehyde in acid solution gives the corresponding aldehydes in good yields [139,141].

Much work has been devoted in the past to the study of the electrochemical reduction of carbon-nitrogen multiple bonds to form amine functions [6,7,9]. In more recent work, Lund examined the polarographic and cpe behavior of a number of ketimines, anils, oximes, azines, phenylhydrazones, and semicarbazones, and concluded that the polarographic reduction of R (R')C=N(Y)R" deponds on the nature of the group Y [142]. If Y is connected to the nitrogen via a carbon atom, the reaction is a two-electron reduction with formation of the saturated compound. On the other hand, if Y is connected to the nitrogen *via* a nitrogen or oxygen atom, the over-all reaction in *acidic* medium is a four-electron reduction of the protonated form, resulting in saturation of the double bond and splitting of the N-Y bond. It also was demonstrated that the splitting of the N-Y bond probably occurs before the reduction of the carbon-nitrogen double bond:

$$R(R')C=N(Y)R'' + H^+ \rightleftharpoons R(R')C=N(Y)R''H^+ \tag{72}$$

$$[R(R')C=N(Y)R'']H^+ + 2e^- + 2H^+ \longrightarrow [R(R')C=NH]H^+ + H\text{-}Y\text{-}R'' \tag{73}$$

$$[R(R')C=NH]H^+ + 2e^- + 2H^+ \longrightarrow R(R')CH\text{-}NH_3^+ \tag{74}$$

This reaction scheme was later confirmed for the reduction of oximes in acidic solution by isolation of the intermediate ketimine in two cases [143]. However, in *alkaline* medium, reduction of the C=N bond in oximes and semicarbazones preceds splitting of the N-Y bond, so that hydroxylamines and hydrazines may be obtained [144]. The stereochemistry of the electrochemical and chemical (H_2/Pt, Na/EtOH) reduction of several imines to amines has been compared recently by Fry and Reed [144b].

Cathodic reduction of camphor and norcamphor oxime [144a] appears to be cases where the stereochemistry is controlled by attack by the electrode from the least hindered side of the molecule:

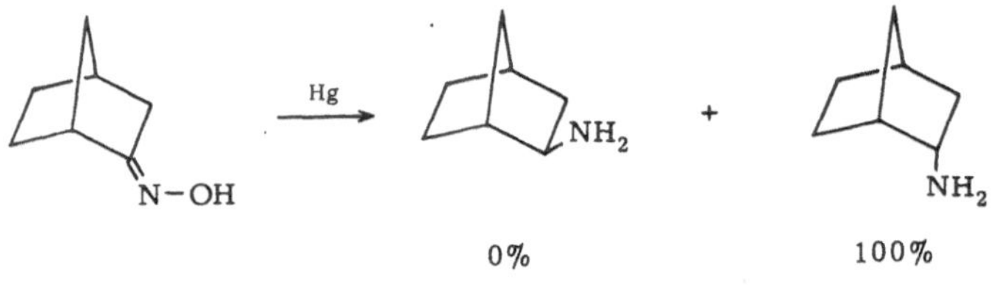

$$\tag{75}$$

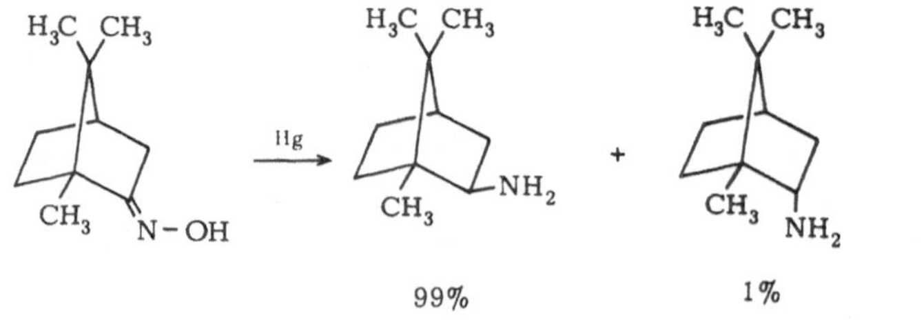

0%   100%

$$\tag{76}$$

99%   1%

The reduction of the above-mentioned types of acyclic C=N-bond containing compounds was later used as model reactions for the cathodic reduction of analogous heteroaromatic and heteroethylenic compounds, such as substituted pyridazinones, cinnolines, phthalazinones, phthalazines, 2,3-benzoxazinones, and benzotriazoles. This work has been summarized in a recent article [145], in which also the influence of protonation, tautomerism, and proton participation in intermediate chemical steps was discussed. Generally, the cyclic compounds behave analogously to the corresponding acyclic models, provided these are chosen with care.

The carbon-nitrogen triple bond in nitriles is reducible to an aminomethyl group in acidic media at lead or mercury electrodes [9], whereas in neutral or alkaline medium cleavage of the C-CN bond takes place [146-148]. This duality of mechanism has been studied in great detail for 2- and 4-cyanopyridine [146] and it was demonstrated that both pH and electrode potential is of importance in determining the mechanism.

Among all functional groups, the nitro group is one of the most easily reducible ones. When present in a molecule, the nitro group in almost all cases is the point of attack in cathodic reduction, leading to compounds containing the hydroxylamine, amine, azoxy, azo, or hydrazo function, depending on the compound and reaction conditions employed [4-7,9]. High product selectivity can normally be achieved withouth difficulty by proper choice of electrode material and electrolyte pH.

More recent work has shown that the cathodic reduction of aliphatic nitro compounds in acidic medium proceeds through the nitroso stage [149] to give the hydroxylamine in a four-electron reaction:

$$\begin{bmatrix} R' \\ | \\ R\text{-}C\text{-}NO_2 \\ | \\ R'' \end{bmatrix} H^+ \xrightarrow{2e^-,\ 2H^+} \begin{bmatrix} R' \\ | \\ R\text{-}C\text{-}NO \\ | \\ R'' \end{bmatrix} H^+ \qquad (77)$$

$$\begin{bmatrix} R' \\ | \\ R\text{-}C\text{-}NO \\ | \\ R'' \end{bmatrix} H^+ \xrightarrow{2e^-,\ 2H^+} \begin{array}{c} R' \\ | \\ R\text{-}C\text{-}NH_2^+OH \\ | \\ R'' \end{array} \qquad (78)$$

The nitroso compound could be directly observed in the reduction of t-nitrobutane at potentials less negative than -1.0 V (SCE); for primary and secondary nitroalkanes, the isolation of carbonyl compounds from the reduction constitutes indirect evidence:

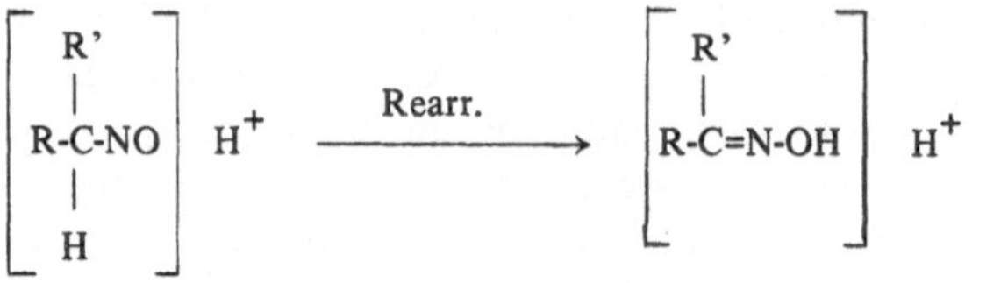

$$\begin{bmatrix} \text{R'} \\ | \\ \text{R-C-NO} \\ | \\ \text{H} \end{bmatrix} \text{H}^+ \xrightarrow{\text{Rearr.}} \begin{bmatrix} \text{R'} \\ | \\ \text{R-C=N-OH} \end{bmatrix} \text{H}^+ \qquad (79)$$

$$\begin{bmatrix} \text{R'} \\ | \\ \text{R-C=N-OH} \end{bmatrix} \text{H}^+ \xrightarrow{\text{H}_2\text{O}} \text{RCOR'} + \text{NH}_3^+\text{OH} \qquad (80)$$

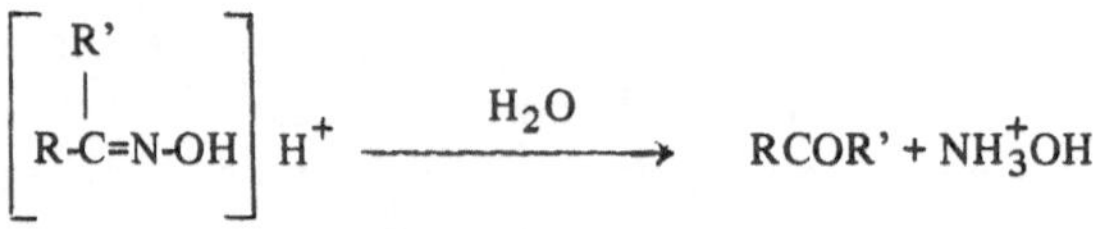

The formation of alkylamines is also indirect evidence for the intervention of nitroso compounds, since the hydroxylamine is not reducible to the amine in acidic medium. Therefore, the amine must be formed by reduction of the oxime formed in the rearrangement of the nitroso compounds (Eqs. (79) and (81)).

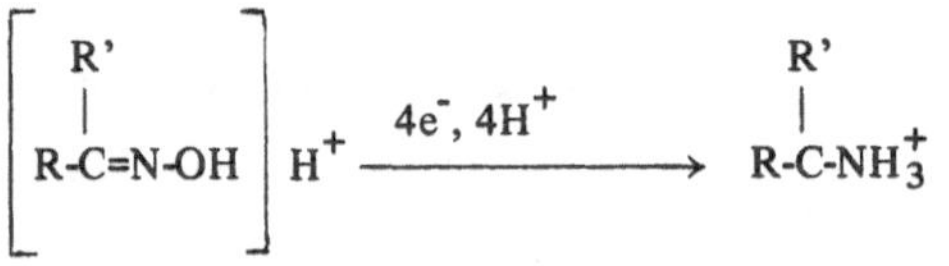

$$\begin{bmatrix} \text{R'} \\ | \\ \text{R-C=N-OH} \end{bmatrix} \text{H}^+ \xrightarrow{4\text{e}^-, 4\text{H}^+} \begin{array}{c} \text{R'} \\ | \\ \text{R-C-NH}_3^+ \end{array} \qquad (81)$$

Thus, the important factor in determining whether the nitroalkane will be reduced to a hydroxylamine or an amine is the temperature, since increased temperatures will speed up the rearrangement reaction (79) and favor the formation of the amine.

In anhydrous acetonitrile, the cathodic reduction of aliphatic t-nitro compounds [150] follows a different mechanism, in that the first step leads to the formation of an anion radical [151], which subsequently is cleaved to form nitrite ion and a neutral radical:

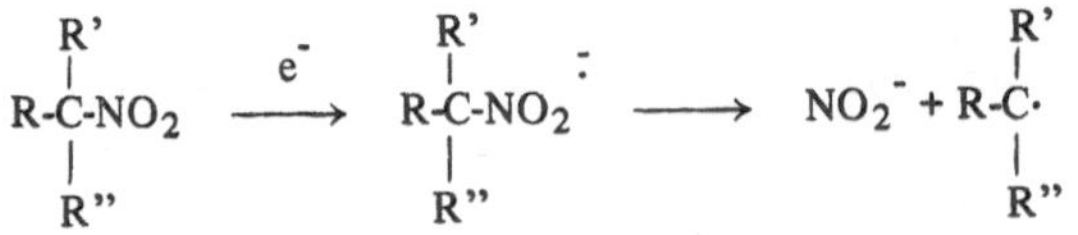

$$\begin{array}{c} \text{R'} \\ | \\ \text{R-C-NO}_2 \\ | \\ \text{R''} \end{array} \xrightarrow{\text{e}^-} \begin{array}{c} \text{R'} \\ | \\ \text{R-C-NO}_2^{\cdot\,-} \\ | \\ \text{R''} \end{array} \longrightarrow \text{NO}_2^- + \begin{array}{c} \text{R'} \\ | \\ \text{R-C}\cdot \\ | \\ \text{R''} \end{array} \qquad (82)$$

Aromatic o-dinitro compounds [152] can be selectively reduced in acidic medium to either an o-nitrophenylhydroxylamine (cpe at less negative potentials) or an o-phenylenediamine (cpe at more negative potentials) as for example in the reduction of dinitrobenzene itself:

9. Electrochemical Substitution

9.1. Anodic Substitution

Anodic substitution denotes those electrochemical processes in which the group X, mostly hydrogen, of a substrate RX is replaced anodically by a substituent Nu, *e.g.*, OR, OCOR, NHCOR.

$$RX + Nu^- \xrightarrow{\text{-2e}} R\text{-}Nu + X^+$$

Mechanistically, these reactions can be divided into three classes:

a) *Radical substitution of hydrogen by anodically generated radicals Nu* (Eq. (89)):

$$Nu^- \xrightarrow{\text{- e}} Nu^{\cdot}$$

$$Nu^{\cdot} + RH \longrightarrow R^{\cdot} + HNu \qquad\qquad (89)$$

$$R^{\cdot} + Nu^{\cdot} \longrightarrow RNu$$

b) *Substitution of the carboxylate group by Nu$^-$ in the Kolbe electrolysis of carboxylates* (Eq. (90)):

$$RCO_2^- + Nu^- \xrightarrow{\text{-2e, -CO}_2} R\text{-}Nu \qquad\qquad (90)$$

c) *Nucleophilic substitution of an anodically generated radical cation* (Eq. (91)):

$$RH \xrightarrow{\text{-e}} RH^{+\cdot} \xrightarrow{Nu^-} RHNu^{\cdot} \xrightarrow{\text{-e/-H}^+} R\text{-}Nu$$

The preparative scope of anodic substitution has not been systematically studied yet. Its synthetic potential can be judged from the possibilities i) to utilize the class of nucleophiles for radical substitution processes and ii) to join two reagents of equal polarity (b, c) by preceding oxidation, a reaction that cannot be matched by conventional synthesis. Thus the choice of reagents for radical substitution is considerably broadened and new combinations of synthetic building units are offered.

The literature on anodic substitution up to 1961 has been reviewed by Tomilov [16].

a) Radical Substitution of Hydrogen by Anodically Generated Radicals

The electrochemical fluorination of the C-H bond has been intensively studied, because it offers a convenient way to prepare perfluorinated organic compounds. Electrochemical chlorination, bromination and iodination have received less interest, because these reactions can be performed − in contrast to fluorination − without difficulty by conventional methods. Schmeisser and Sartori [175] and Forche[176] have reviewed the literature on electrochemical fluorination and the reader is referred to this articles for details (see also Sect. 15.4, Ref. 168, 169). Fluorination is normally conducted in anhydrous HF/LiF (NaF, KF) in vessels of iron, copper or polyvinyl chloride without a diaphragm at iron, copper or nickel cathodes and nickel anodes. Normally perfluorinated products are obtained in only 20 to 30 % material yield. The greater part of the substrate is disintegrated to smaller fragments by C-C bond cleavage, presumably by elementary fluorine. Fluorination without C-C bond destruction seems to depend on the formation of a phase layer of nickel fluoride at the anode surface [177]. The exact mechanism of electrofluorination is not known, which makes classification of this reaction as a radical process somewhat arbitrary. Elementary fluorine is not an intermediate, but it is possible that higher valence nickel fluorides play an intermediate role. Table 3 illustrates the preparative application of electrofluorination in some representative examples.

Table 3. *Electrofluorination of organic compounds*

Substrate	Product	Ref.
Octane	$CF_4(10.4c^{a)})$, $C_2F_6(3c)$, $C_8F_{18}(12.5$ c$)$	[178]
CH_3Cl	CF_4, $CClF_3$, CHF_3, $CHClF_2$, CH_2F_2, CH_2ClF, CH_3F	[179]
$CClBr_2CCl_3$	CF_2ClCCl_3 (70c, $90m^{b)}$)	[180]
$CH_2=CHCl$	All possible combinations (20–70 m)	[181]
CH_3COF	CF_3COOH (71 m)	[182]
Malonic, succinic, maleic diesters	Perfluorinated dicarboxylic acids (10–15 m)	[183]
$CH_3(CH_2)_nCOOH$	$CF_3(CF_2)_nCOOH$, substituted perfluorinated tetrahydrofurans	[184]
Subst. tetrahydrofurans	Subst. perfluorinated tetrahydrofurans	[185]
CH_3SO_2F	CF_3SO_2F (96m)	[187]
Tetrahydropyrane	Decafluorotetrahydropyrane (34.5 m)	[186]
$C_8H_{17}SO_2F$	$C_8F_{17}SO_2F$ (25 m)	[187]
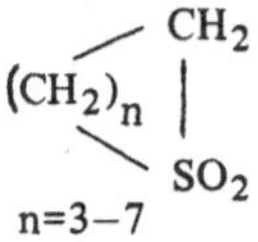 $(CH_2)_n \begin{smallmatrix} \diagup CH_2 \\[2pt] \diagdown SO_2 \end{smallmatrix}$ n=3–7	$CF_3(CF_2)_nSO_2F$	[188]
t- and *sec*-Amines	Perfluoroamines	[189]
Pyridine	$C_5F_{12}(77$ c$)$, $NF_3(20$ c$)$	[178]
N-Methylpiperidine	N-Trifluoromethyldecafluoropiperidine	[190]

a) Current yield %, b) Material yield %.

Suitable hydrogens, *e.g.*

allylic-H, $>$CHOR, -CON(CH$_3$)$_2$, $>$SiH, R,R,R-CH, $\overset{\displaystyle O}{\overset{\displaystyle \|}{-CH}}$, $\overset{\displaystyle O}{\overset{\displaystyle \|}{-CCH-}}$

can be conveniently substituted by -OCH$_3$, -OAc, or -OH on electrolysis of the substrate in the corresponding solvent. The ethers RR'CHOR" dioxane, tetrahydrofurane, tetrahydrothiophene, and cyclopropylcarbinyl methyl ether can be converted in 8% to 28% yield to the corresponding acetals RR'COR"-OCH$_3$ on electrolysis in methanol-CH$_3$ONa, -NH$_4$NO$_3$ or -Et$_4$NClO$_4$ [191]. Cyclohexene, 1-octene or 1,5-cyclooctadiene form in methanol-, acetic acid- or acetonitrile-NH$_4$NO$_3$ at carbon electrodes in 14 to 55% yield the corresponding alkenyl methyl ethers, acetates, or acetamides [192]. These substitutions are rationalized by the following scheme (Eq. (92)):

$$\text{supporting electrolyte} \quad \xrightarrow{-e} \quad S \cdot$$

$$ROCH_2R' \xrightarrow{S \cdot} RO\overset{\cdot}{C}H\text{-}R' \xrightarrow{-e} RO\overset{+}{C}HR' \xrightarrow{R''OH} RO\overset{\cdot}{C}HR' \quad (92)$$
$$\qquad\qquad\qquad\qquad\qquad\qquad\qquad\qquad\qquad\qquad\qquad \underset{OR''}{|}$$

The supporting electrolyte is anodically oxidized to radicals, which abstract α-hydrogen from the ether to form alkoxymethyl radicals that are subsequently oxidized to cations. An analogous route leads to alkenyl derivatives.

A radical substitution path is assumed for the alkoxylation or acetoxylation of the methyl group in N,N-dimethylbenzamide, -acetamide or -formamide to the corresponding N-alkoxy- or N-acetoxymethyl derivatives [193]. An alternative path via radical cations has been proposed for the acyloxylation of amides on the basis of polarographic measurements and controlled potential electrolysis (Eq. (93)) [48]

$$RC\text{-}N(CH_3)_2 \xrightarrow{-e} R\overset{+\cdot}{C}N(CH_3)_2 \xrightarrow{-H^+/-e^-} R\overset{}{C}N\overset{CH_2^+}{\underset{CH_3}{|}} \xrightarrow{HOAc} R\overset{}{C}N\overset{CH_2OAc}{\underset{CH_3}{|}} \quad (93)$$
$$\underset{O}{\|} \qquad\qquad \underset{O}{\|} \qquad\qquad \underset{O}{\|} \qquad\qquad \underset{O}{\|}$$

Anodic methoxylation of cyclohexyl isocyanide yields a mixture of six products whose formation is thought to be initiated by methoxy radicals [194]. Cholesteryl acetate dibromide has been specifically hydroxylated at C25 in 85 to 93% yield at a lead dioxide anode [195]. Silyl ethers have been prepared in 60 to 95 % yield by electrolysis of trialkylsilanes at a platinum electrode in ROH- Me_4NI, - $NaNO_3$, or - Et_4NBr [196]. Both reactions presumably occur via the radical path (89).

Methyl radicals generated by electrolysis of acetic acid can be used for radical substitution of aldehydic hydrogen or α-hydrogen in ketones. As an example, enanthaldehyde yields 2-octanone, 2-octanone 3-methyl-2-octanone, and benzaldehyde acetophenone on electrolysis in aqueous acetic acid [197].

b) Substitution of the Carboxylate Group by Nu⁻ in the Kolbe Electrolysis of Carboxylates

The term Kolbe electrolysis is sometimes exclusively used to denote the formation of alkyl dimers R-R, path a) in Eq. (94) (see Sect. 12.1), while the reaction leading to substituted alkyls, R-Nu, via carbonium ions as intermediates, path b) in Eq. (94), is named non-Kolbe electrolysis, abnormal Kolbe electrolysis, or

Hofer-Moest reaction. We do not attempt this discrimination as the electrochemical decarboxylation of carboxylates is always a blend of both pathways.

$$RCO_2^- \xrightarrow{-e} CO_2 + R^{\cdot} \begin{cases} a) \nearrow R\text{-}R \quad \text{dimer} \\ \\ b) \searrow R^+ \xrightarrow{Nu^-} R\text{-}Nu \\ \qquad\qquad\qquad\quad \text{substitution product} \end{cases} \qquad\qquad (94)$$

Eberson has comprehensively reviewed the mechanism of electrochemical decarboxylation [14]. An extensive list of Kolbe electrolyses leading to predominant formation of substitution products has been compiled by Weinberg [10]. Therefore this mode of anodic substitution, though comprehending a wealth of reactions, can be treated briefly and the reader is referred to these articles for more details.

The substitution path b) (Eq. (94)), is favored by the following experimental conditions: low current density, graphite as anode material, alkaline medium, water or water-pyridine as solvent, and admixture of foreign ions: *e.g.*, bicarbonate, sulfate, perchlorate, dihydrogen phosphate, Pb^{2+}, Mn^{2+}, Cu^{2+}, Fe^{2+}, Co^{2+}. The carbonium ion path b) can furthermore be expected for carboxylates RR'CHOO⁻ with α-substituents R' such as alkyl, phenyl [198], hydroxy, halogen [199], amino, or alkoxy. These substituents facilitate oxidation of the intermediate alkyl radical $R^{\cdot}$ to the carbonium ion R^+. Product formation occurs *via* carbonium ions and not, as is also conceivable, *via* mixed coupling of $R^{\cdot}$ with $Nu^{\cdot}$. In most cases the product RNu, *e.g.*, an ether, alcohol or ester, contains a rearranged carbon skeleton, typical for a carbonium ion intermediate, while the Kolbe dimers R-R, originating from alkyl radicals, contain unrearranged alkyl groups. Besides rearrangement the intermediate cation R^+ can undergo 1,3-elimination to cyclopropanes (path a, Eq. (95)) [82,200] or transannular elimination in 9- or 10 -membered rings (path b) [200]. The cations form olefins by E1 elimination (path c) or solvolyze to esters, alcohols (path d), ethers, or acetamides, depending on the nature of the SSE.

Very similar product mixtures are obtained from electrochemical decarboxylations or deoxidations [201] and deaminations [202] of the corresponding alcohols or amines. Anodic oxidation of cyclobutanecarboxylic acid affords in 30% yield a mixture of cyclopropylcarbinol, cyclobutanol and allylcarbinol identical in composition with that obtained from deamination of cyclobutylamine [203]. Electrolysis of *exo-* or *endo*-norbornane-2-carboxylic acid gave *exo*-norbornyl-2-

methylether in 35% to 40% yield [203]. From *exo-* or *endo*-5-norbornene-2-car-
boxylic acid (*19*) 3-methoxy-nortricyclene (*20*) is formed (Eq. (96)) in 56%
yield [203]. In both cases the products correspond exactly to those obtained from
solvolysis.

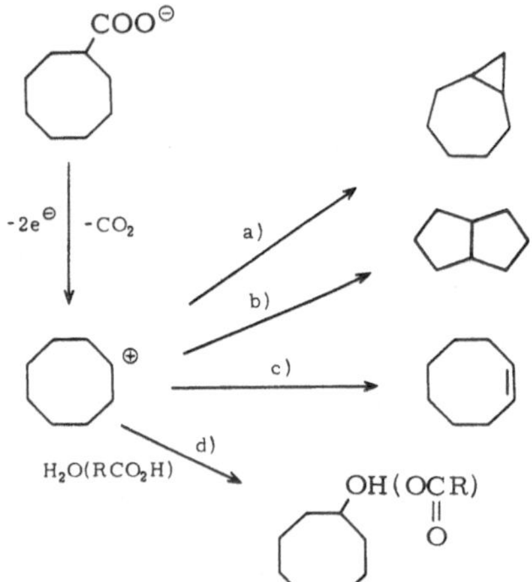

$$(95)$$

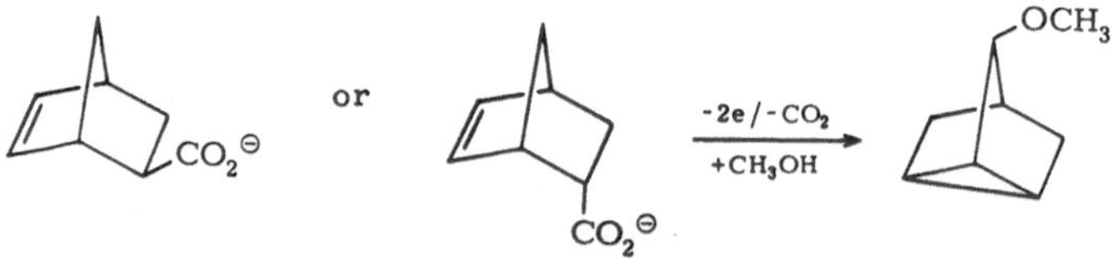

| *19* | *20* | *21* |

$$(96)$$

    The mode of cation formation in electrochemical decarboxylation appears
not to be uniform. Skell [204] found two discrete 1e-steps for oxidation of carboxylates by chronopotentiometry. He attributed the second electron transfer
to oxidation of the alkyl radical (path b, Eq. (94)) as the carboxylate radical
$RCO_2 \cdot$ is to shortlived (τ 10^{-10} sec) to survive for the second oxidation step.
    On the other hand, cation formation by decarboxylation of an acyloxonium
cation RCO_2^+ is supported by the partial stereospecificity observed in the electrolysis of *cis-* and *trans*-bicyclo [3.1.0] hexane-3-carboxylic acid [205] and the
electrocyclic ring opening in the anodic oxidation of 3-methyl-2-phenylcyclopro-
panecarboxylate (*22, 23*) to cyclopropyl methyl ether (*24, 25*) and allylic ethers
(*26, 27*) (Eq. (97)) [206].

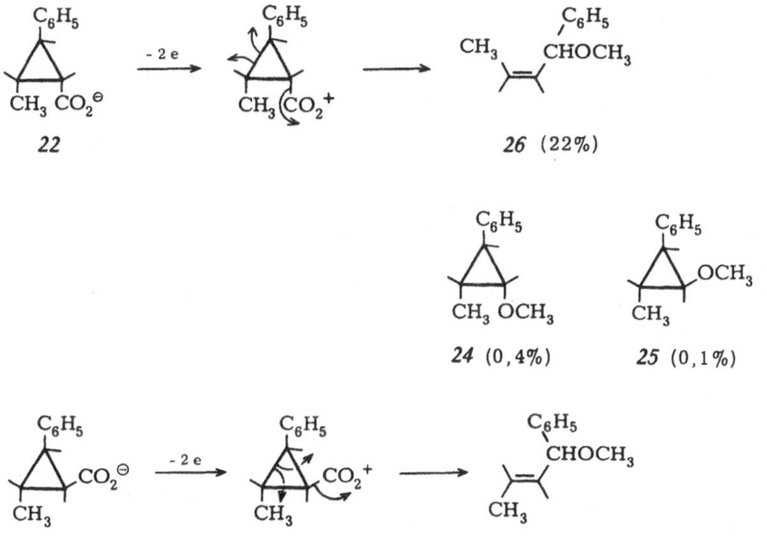

$$
\underset{22}{\overset{\text{C}_6\text{H}_5}{\triangle}\!\!\!\begin{array}{c}\\ \text{CH}_3 \ \text{CO}_2{}^{\ominus}\end{array}} \xrightarrow{-2\,e} \overset{\text{C}_6\text{H}_5}{\triangle}\!\!\!\begin{array}{c}\\ \text{CH}_3 \ \overset{+}{(}\text{CO}_2)\end{array} \longrightarrow \underset{26 \ (22\%)}{\overset{\text{C}_6\text{H}_5}{\text{CH}_3}\!\!\!\begin{array}{c}\\ \text{CHOCH}_3\end{array}}
$$

$$
\underset{24 \ (0,4\%)}{\overset{\text{C}_6\text{H}_5}{\triangle}\!\!\!\begin{array}{c}\\ \text{CH}_3 \ \text{OCH}_3\end{array}} \qquad \underset{25 \ (0,1\%)}{\overset{\text{C}_6\text{H}_5}{\triangle}\!\!\!\begin{array}{c}\\ \text{CH}_3\end{array}\!\!\text{OCH}_3} \tag{97}
$$

$$
\underset{23}{\overset{\text{C}_6\text{H}_5}{\triangle}\!\!\!\begin{array}{c}\\ \text{CH}_3\end{array}\!\!\text{CO}_2{}^{\ominus}} \xrightarrow{-2\,e} \overset{\text{C}_6\text{H}_5}{\triangle}\!\!\!\begin{array}{c}\\ \text{CH}_3\end{array}\!\!\text{CO}_2{}^{+} \longrightarrow \underset{27 \ (6\%)}{\overset{\text{C}_6\text{H}_5}{\text{CH}_3}\!\!\!\begin{array}{c}\\ \text{CHOCH}_3\end{array}}
$$

$$
+ \ 24 \ (9\%) \ + \ 25 \ (2\%)
$$

Anodic decarboxylation produces – as deamination and deoxidation does – a poorly solvated reactive 'hot' carbonium ion [207,14] being less encumbered by solvent molecules than cationic species generated by solvolysis. The 'free' cation rearranges before it becomes encumbered by neighbor nucleophiles in $\sim 10^{-10}$ sec and collapses to products (ether, alcohol, olefin), while in the corresponding tosylate systems solvolysis involving encumbered cations occurs without significant rearrangement [207]. The electrogenerated cation can be trapped more efficiently before rearrangement by raising the nucleophilicity of the solvent, *e.g.*, by increasing the OH$^-$ concentration [208].

Table 4 gives some illustrative examples of the preparative potential of anodic substitution *via* anodic decarboxylation of carboxylates.

Table 4. *Anodic substitution of carboxylates*

Carboxylic acid	Product	Ref.
Rearrangements		
1-Hydroxycyclohexane-acetic acid	Cycloheptanone (45–53%)	203)

Table 4 (continued)

Carboxylic acid	Product	Ref.
1-Hydroxycycloheptane- acetic acid	Cyclooctanone	203)

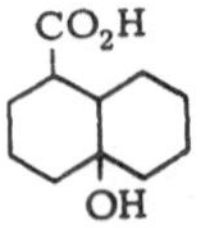

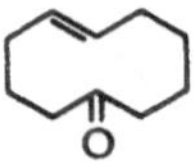

(44%) 209)

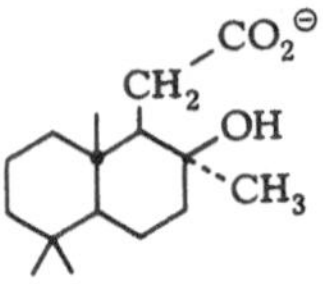

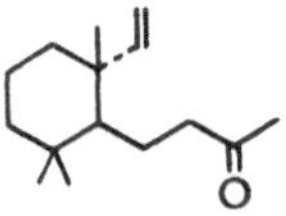

210)

| 2,2-Dimethylcyclopro-
panecarboxylate | Allyl methyl ether, allyl cyclopropanecarboxylates | 211) |

Intramolecular substitution

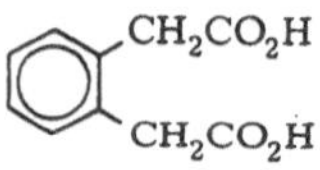

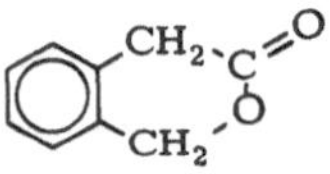

212)

| *o*-Benzoylbenzoic
acid | Fluorenone (9%), 3-arylphthalide (20%), arylmethyl
phthalate (27%) | 213) |
| Polymethacrylic acid | Lactone 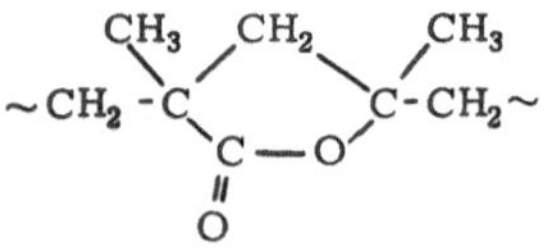 | 214) |

(35–40%)

Alkoxylation:

$$RCO_2^- \xrightarrow[-CO_2]{-2e^-/R'O^-} ROR'$$

Diphenylacetic acid	Benzhydryl methyl ether (35–80%)	215)
α-Ethoxyphenyl- acetic acid	Benzaldehyde methylethylacetal (71%)	216)
α-Methoxydiphenyl- acetic acid	Benzophenone dimethylacetal (74%)	216)
N-Benzoylglycine	N-Methoxymethylbenzamide (61%)	217)

Table 4 (continued)

Carboxylic acid	Product	Ref.
N-Benzoyl-d, 1-α-alanine	N-1'-Methoxyethylbenzamide (91%)	[217]
1-Azabicyclo [2.2.2]-octane-2-carboxylic acid	2-Methoxy-1-azabicyclo [2.2.2] octane (43%)	[218]

Nitrates: $RCO_2^- \xrightarrow[-CO_2]{NO_3^-/-2e^-} R'ONO_2$

Sodium adipate (sodium nitrate, H_2O)	1,2 Butanediol dinitrate, 1,2,3-butanetriol trinitrate, 1,2,4-butantriol dinitrate	[219]
Potassium malonate (potassium nitrate, H_2O)	Glykol dinitrate, 1,4-butandiol dinitrate	[220]
Potassium succinate (potassium nitrate, H_2O)	Glykol dinitrate, 1,4-butandiol dinitrate	[221]

Acetamides: $RCO_2^- \xrightarrow[CH_3CN/H_2O]{-2e^-/-CO_2} R-\overset{\displaystyle H}{\underset{\displaystyle O}{N}}\overset{}{\underset{\|}{C}}CH_3$

Trimethylacetic acid	N-t-Butylacetamide (40%)	[222]
t-Butylcyanoacetic acid	N-(t-Butylacetyl-t-butylglycinonitrile) (31%)	[223]
Valeric acid	2-Butylacetamide (60%)	[224]
$CH_3(CH_2)_nCO_2H$ (n=0–3)	$CH_3\overset{O}{\overset{\|}{C}}-\underset{(CH_2)_nCH_3}{N}-\overset{O}{\overset{\|}{C}}-(CH_2)_nCH_3$, $CH_3\overset{O}{\overset{\|}{C}}-NH-\overset{O}{\overset{\|}{C}}-(CH_2)_nCH_3$	[225]
Cyclohexanecarboxylic acid	$CH_3CON-\overset{O}{\overset{\|}{C}}_6H_{11}(-C\ C_6H_{11})$ (32%)	[225]

c) Nucleophilic Substitution of an Anodically Generated Radical Cation

Nuclear and side chain substitution in aromatics or substitution of α-hydrogen in alkylamines is — in most cases — best rationalized by postulating radical cations as intermediates. For anodic nuclear substitution of aromatics, especially for acyloxylation, cyanation or bromination a $EC_N EC_B$ [38]-mechanism is assumed [37,49,50,226,227]: 1e-oxidation of the aromatic to the radical cation *28*, which reacts with a nucleophile Nu^-, *e.g.*, acetate, cyanide, alkoxide, followed by a second electron transfer and deprotonation (Eq. (98)):

$$\tag{98}$$

An alternative radical process (Eq. (99)):

$$Nu^- \longrightarrow Nu^\cdot$$

$$RH + Nu^\cdot \longrightarrow RHNu^\cdot \tag{99}$$

$$RHNu^\cdot + Nu^\cdot \longrightarrow RNu + HNu$$

has been suggested by several authors [228-230]. However, this sequence is clearly ruled out for acetoxylation [48], cyanation [57,60,227], bromination [231] and pyridination [37]. In aromatic acetoxylation the acetates are obtained in fair yields by cpe well below the discharge potential of the acetate ion. Attempted bromination or cyanation by cpe at the discharge potentials of bromide or cyanide ion yields no substitution products, while fair to good yields of bromides or nitriles are obtained at higher potentials corresponding to the oxidation potentials of the aromatics. Adams *et al.* [37] unequivocally demonstrated an ECE-mechanism for the pyridination of 9,10-diphenylanthracene. Using the rotating disc electrode technique rates for the reaction of the radical cation with various nucleophiles could be determined. The influence of steric hindrance in the reaction of the intermediate cation radical with nucleophiles was studied by Parker and Eberson [231a]. Pyridination of the radical cation of 9,10-diphenylanthracene or 1,4-dimethoxybenzene occurs 30 to 40 times slower with 2,6-lutidine than with the sterically less hindered 3,5-lutidine, although both have roughly the same basicity. Methoxylation of aromatics has always been conducted by uncontrolled potential electrolysis. Therefore, the suggested intermediacy of methoxy radicals cannot be positively ruled out in aromatic methoxylation. However, in the methoxylation of anisole the high ratio of $o+p/m$-substitution product: 97/3 [232]), being close to that for acetoxylation of anisole 96,5/3,5 [226], reflects the charge distribution in the radical cation *28* and disfavors the radical substitution path (Eq. (99)) for which a lower $o+p/m$-product ratio would be expected.

Table 5 summarizes some examples of anodic aromatic nuclear substitution to give some illustration of synthetic applications. For further details the reader is referred to Ref. 10, 25.

Table 5. *Anodic nuclear substitution of aromatics*

Aromatic compound	Product	Ref.
Acyloxylation:	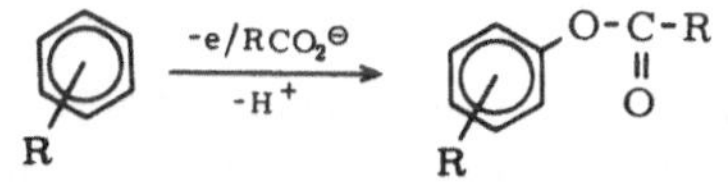	
Toluene (HOAc, NaOAc)	Monoacetoxytoluenes (o:m:p = 43:11:45) benzyl acetate	226)
t-Butylbenzene	o-,m-,p-t- Butylphenyl acetates (30:24:45)	226)
Anisole (HOAc, NaOAc)	o- and p-Acetoxyanisole (40%)	48)
Anisole (CH$_3$CN, Et$_3$N, benzoic acid)	o-, p-Methoxyphenyl benzoate (60%)	232a)
Naphthalene (HOAc, NaOAc)	1-Acetoxynaphthalene, 1- and 2-methyl-naphthalene	48)
Cumene, toluene, ethylbenzene (CH$_3$CN, Et$_3$N, benzoic acid	Corresponding benzoyloxyaromatics	233)

Intramolecular Acyloxylation:

3,3-Diphenylacrylic acid (CH$_3$OH, KOH)	4-Phenylcoumarin (58%)	234)
Biphenyl-2-carboxylic acid	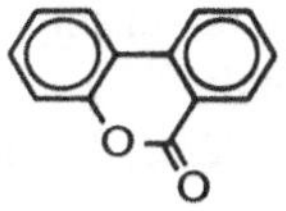	48)

Cyanation, thiocyanation

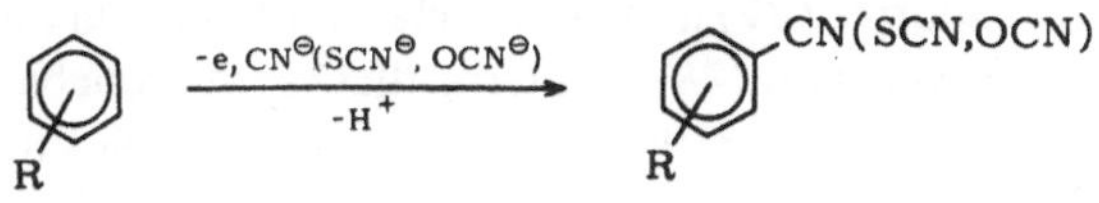

Anisole, toluene, ethylbenzene (NaCN, MeOH)	o- and p-Nitriles (low current yields: 1–12%)	227,235) 58)

69

Table 5 (continued)

Aromatic	Product	Ref.
Toluene, mesitylene	Exclusive nuclear cyanation, selective in o- and p-position	57)
Diphenylacetylene	p-Cyanodiphenylacetylene	236)
1,2-Dimethoxybenzene (Et_4NCN, CH_3CN)	o-Cyanoanisole (94%)	60)
1,4-Dimethoxybenzene (Et_4NCN, CH_3CN)	p-Cyanoanisole (95%)	60)
Anthracene (Et_4NCN, CH_3CN)	9,10-Dicyanoanthracene (54%)	60)
Biphenyl (Et_4NCN, CH_3CN)	4-Cyanobiphenyl (20%)	60)
Dimethylaniline (NH_4SCN, HCl, EtOH, H_2O)	N, N-Dimethyl-4-thiocyanoaniline (85%)	237)
Anisole (MeOH, KCNO)	o-, p- Methoxyphenylcarbamate ($<$20%)	61)
Bromination:		
Anthracene (Et_4NBr, CH_3CN)	9-Bromoanthracene (48%, cpe 1,2 V Ag/Ag^+) 9,10-Dibromoanthracene (19%, cpe 1,6 V Ag/Ag^+)	231)
Naphthalene (Et_4NBr, CH_3CN)	1-Bromonaphthalene (70%, 1,35 V) 1,4-Dibromonaphthalene (71%, 2,0 V)	231)
Sulfonation:		
Hydroquinone (Na_2SO_3, H_2O)	Mono- and disulfonated hydroquinone	238)

Phenol has been hydroxylated nearly quantitatively to hydroquinone [239]. Most alkoxylations or hydroxylations of aromatics however either lead to anodic additicn products (see Sect. 10.1) or to side chain substitution (see below). Specific side-chain hydroxylation is difficult to achieve because the alcohols formed as primary products are further oxidized to aldehydes, ketones and/or carboxylic acids.

Anodic iodination [240] involves an iodonium intermediate, probably N-iodo-acetonitrilium perchlorate (29) undergoing electrophilic aromatic substitition (Eq. (100)). A radical cation (28) as intermediate is improbable in this case. Electrolysis of iodine and aromatics in $CH_3CN/LiClO_4$ yields the corresponding

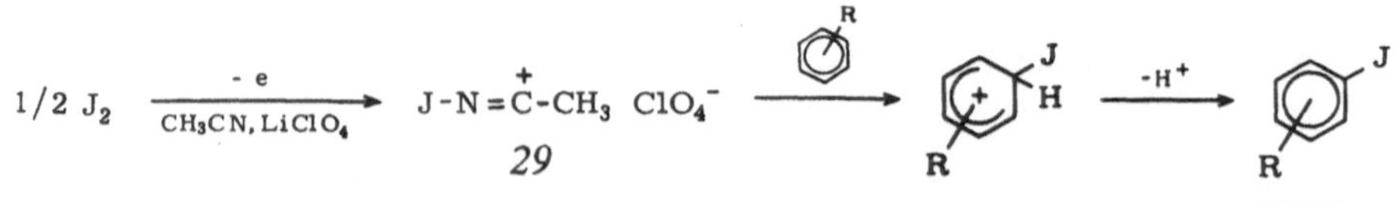

$$1/2\ J_2 \xrightarrow[CH_3CN,\,LiClO_4]{-\,e} J\text{-}N{=}\overset{+}{C}\text{-}CH_3\ ClO_4^- \longrightarrow \longrightarrow \xrightarrow{-H^+}$$

29

$$(100)$$

iodoaromatics, *e.g.* benzene (11%), toluene (34%), xylene (50%), mesitylene (73%), anisole (19%) and triphenylmethane (15%), besides side chain substitution products in some cases. Side-chain substitution is totally suppressed and the yields in iodo-aromatics are raised to 80–100% by the aromatic compound after electrolysis.

Side-chain substitution of aromatics is best rationalized by an EC_BEC_N mechanism *via* a radical cation *30* in Eq. (101) as intermediate [106,226,241–243]. Yet side products of typical radical origin, *e.g.*, bibenzyl in acetoxylation of toluene, have been accounted in favor of a radical chain mechanism (Eq.(99)) [230, 244,245]. An ECE-mechanism however has been clearly demonstrated by cyclic voltammetry for side-chain substitution of pentamethylanisole and *p*-methoxytoluene [241]. Eberson has proposed a modified EC_BEC_N mechanism to account for the formation of radical coupling products [242] (Eq. (101)): The radical cation *30*, the first intermediate, can escape from the electrode surface and loose a proton to form a benzyl radical in the bulk of the solution. This benzyl radical can couple to bibenzyl or abstract hydrogen to form starting material.

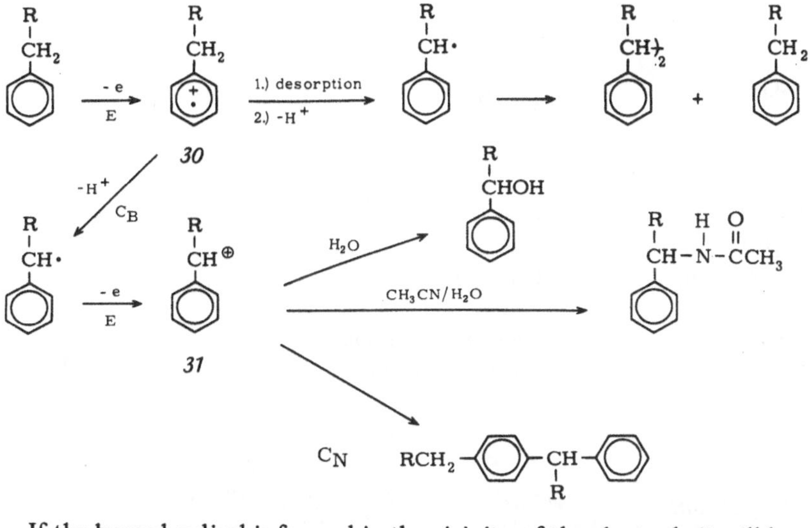

$$(101)$$

If the benzyl radical is formed in the vicinity of the electrode it will be oxidized to the benzyl cation *(31)* $(E_{1/2} \sim +0.9 \text{ V})$ that depending on its reactivity and the nucleophiles present in the solvent will form benzyl alcohol [241,242], N-benzylacetamide [241,242], benzyl nitrate [244] or in the absence of nucleophiles will react with excess aromatic compound to diphenylmethane derivatives; *e.g.*, heptamethyldiphenylmethane or trimethyldiphenylmethane in the electrolysis of durene or *p*-xylene in methylene chloride/*n*-Bu$_4$NClO$_4$ as SSE [78]. Side-chain acetoxylation seems to be favoured by the following SSE's: HOAc/NaClO$_4$, HOAc/Et$_4$NOTs [226], HOAc/Et$_4$NNO$_3$ [244].

The preparative scope of anodic side chain substitution is shown in Table 6.

Table 6. *Anodic side-chain substitution of aromatics*

Aromatic	Product	Ref.
Acetamidation:	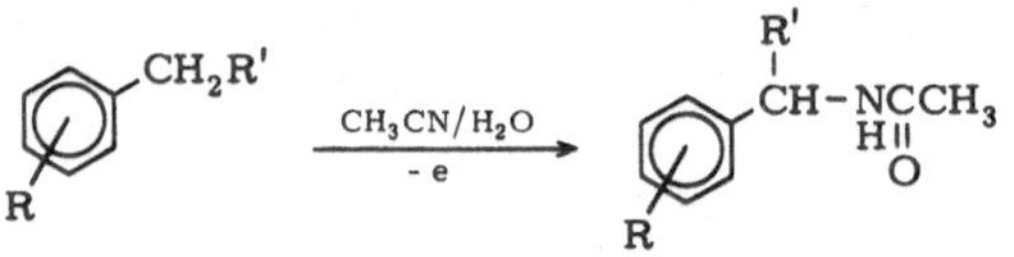 $\text{C}_6\text{H}_3(R)\text{-CH}_2R' \xrightarrow[-e]{\text{CH}_3\text{CN/H}_2\text{O}} \text{C}_6\text{H}_3(R)\text{-CH}(R')\text{-NCCH}_3\,(\text{H})(\overset{\parallel}{O})$	
1,2,4,5-Tetramethyl-benzene (CH_3CN, $NaClO_4$)	N-2,4,5-Trimethylbenzylacetamide (38%)	243)
Hexamethylbenzene $CH_3CN/NaClO_4$	N-Pentamethylbenzylacetamide (42%)	243)
p-Methoxytoluene	N-*p*-Methoxybenzylacetamide	241)
Alkoxylation:	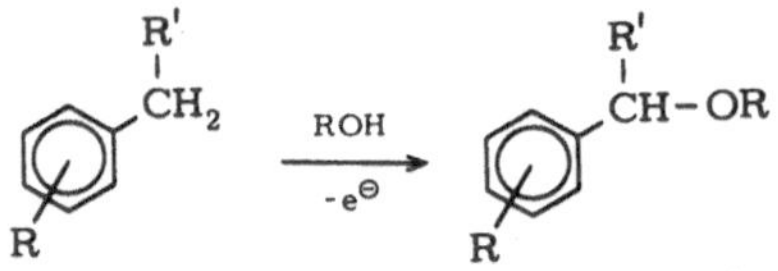 $\text{C}_6\text{H}_3(R)\text{-CH}_2R' \xrightarrow[-e^{\ominus}]{\text{ROH}} \text{C}_6\text{H}_3(R)\text{-CH}(R')\text{-OR}$	
Tetraline (CH_3OH, $NaOCH_3$)	α-Methoxytetraline (22–29%)	228)
Indane (CH_3OH, $NaOCH_3$)	α-Methoxyindane (15%)	228)
4,4'-Dimethoxy-diphenylmethyl ether	4,4'-Dimethoxybenzophenone dimethylacetal (60%)	246)
Acetoxylation:	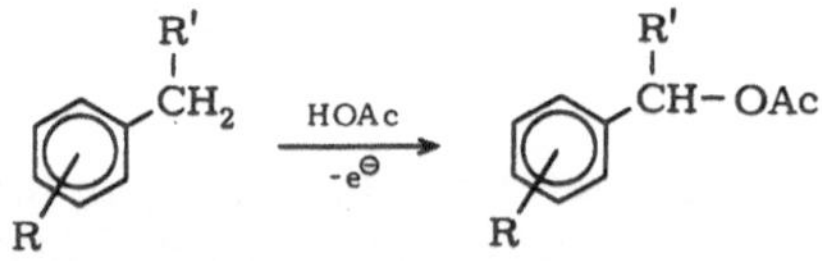 $\text{C}_6\text{H}_3(R)\text{-CH}_2R' \xrightarrow[-e^{\ominus}]{\text{HOAc}} \text{C}_6\text{H}_3(R)\text{-CH}(R')\text{-OAc}$	
Durene (HOAc, NaOAc)	2,4,6-Trimethylbenzyl acetate (46%)	48)
Hexamethylbenzene (HOAc, NaOAc)	Pentamethylbenzyl acetate (57%)	48)

Radical cations are assumed as intermediates in the alkoxylation of alkyl-amines. Alkoxylation of N,N-dimethylbenzylamine (DMB) yields N-α-methoxy-

benzyl-N,N-dimethylamine and N-benzyl-N-methyl-N-methoxymethylamine
(32) in the ratio 1:4 [63,247], while in homogeneous processes the benzyl position is exclusively substituted. This unusual result is rationalized by the following scheme invoking adsorbed intermediates (Eq. (102)):

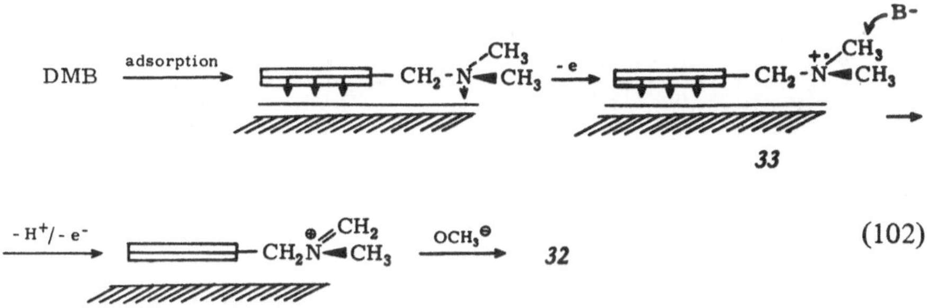

$$(102)$$

In the radical cation 33 the benzyl position is simultaneously deactivated and shielded for attack by the external base, thus accounting for the predominant formation of 32 . Similarly N-benzyl-N-methylethanolamine yields 3-methyl-2-phenyloxazolidine and 3-benzyloxazolidine in the ratio 1:1 by intramolecular substitution (Eq. (103)) [62].

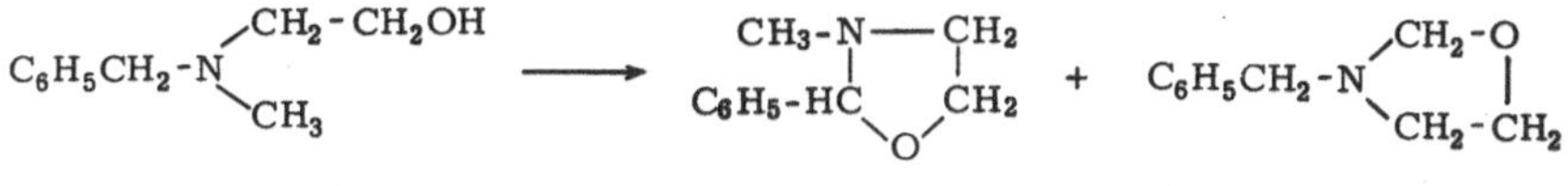

$$(103)$$

N,N-Dimethylaniline (DMA) forms N-methoxymethyl-N-methylaniline (34) and N,N-bis (methoxymethyl)-aniline (35) in methanol-NaOCH$_3$ as SSE in an $EC_B EC_N$ sequence, while in neutral medium with NH_4NO_3/MeOH as SSE N,N,N',N'-tetramethylbenzidine (36) is formed via an EC-mechanism by dimerization of the intermediate radical cation 37 (Eq. (104)) [248]:

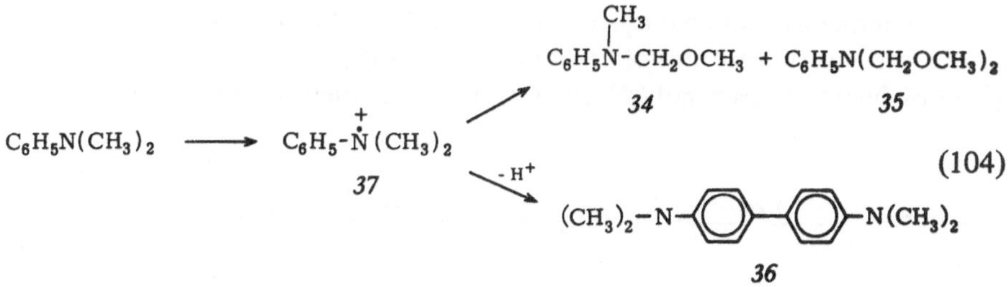

$$(104)$$

2,4,6-Tri-*t*-butylaniline is oxidized in acetonitrile to a radical cation. This reacts with pyridine to form 60% 6,8-di-*t*-butylpyrido[1.2 a] benzimidazole (*38*) by *t*-butyl-transfer to acetonitrile and subsequent dehydrogenation and 25% of the amidine *39*.

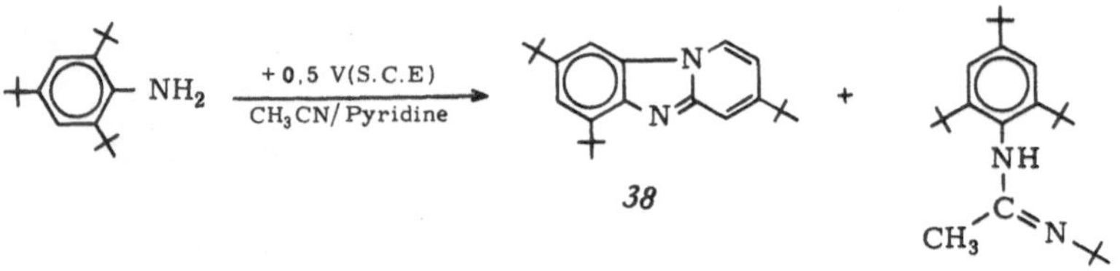

A similar sequence appears to be involved in the formation of octafluorophenazine (*40*) in the oxidation of pentafluoroaniline in HOAc/H$_2$O [250].

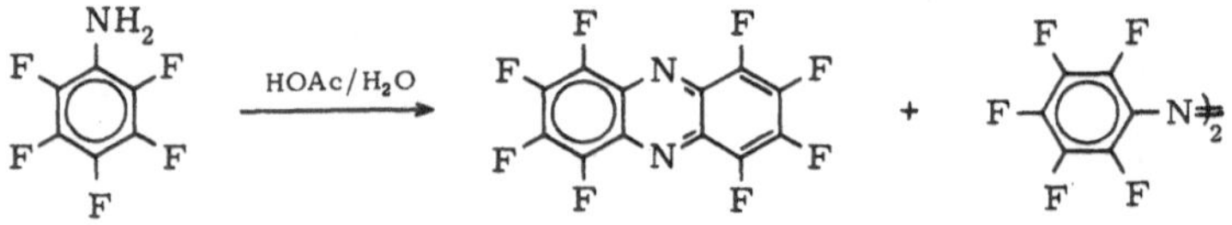

9.2. Cathodic substitution

In cathodic substitution anions are generated from neutral substrates E at the cathode, which react with alkyl halogenides present in the electrolyte to form substitution products E-R according to the scheme (Eq. (105)):

$$E + e^- \longrightarrow E^- \xrightarrow{RX} E\text{-}R + X^- \tag{105}$$

This reaction has been applied both to synthesis and as a mechanistic criterion for the intermediacy of anions, *e.g.*, in the reduction of carbonyl compounds in aprotic solvents. Electrolysis of benzophenone in the presence of ethyl bromide yields diphenylethylcarbinol (*41*, Eq. (106)) [251]. When anthraquinone

$$(C_6H_5)_2C{=}O \xrightarrow{\text{+ 2e}} (C_6H_5)_2\overset{..}{C}\text{-}\overset{..}{O} \xrightarrow{EtBr/H^+} (C_6H_5)_2\overset{\overset{\displaystyle Et}{|}}{C}OH \tag{106}$$

41

74

is reduced in acetonitrile/ethyl iodide a moderate yield of 9,10-diethoxyanthracene is obtained [252]. Reduction of α-haloketones in aprotic solvents in the presence of alkylating agents yields α-alkylsubstituted ketones [253].

Intramolecular cathodic substitution was used by Rifi [254] to construct a very efficient and powerful synthesis for cyclopropanes or bicyclobutanes starting from 1,3-dibromo compounds. Electrolysis of 1,3-dibromocyclobutane in DMF/LiBr gave 60% bicyclobutane, 20% cyclobutene and 10% cyclobutane. 1,3-Dibromo-1,3-dimethylcyclobutane was converted almost quantitatively to 1,3-dimethyl-bicyclobutane (Eq. (107)).

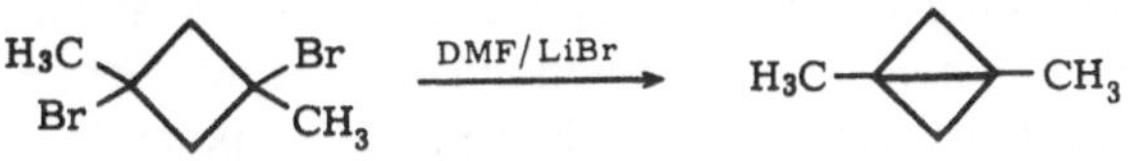

$$\text{(107)}$$

Similarly 1,3-dichlorobicyclobutane (Eq. (108)) and spiropentane (Eq. (109)) were prepared. The tendency for ring closure decreases with increasing ring size. While cyclopropane is formed in 100% yield from 1,3-dibromopropane, only 25% of cyclobutane is obtained from 1,4-dibromobutane and no cyclopentane is formed from 1,5-dibromopentane.

$$\text{(108)}$$

$$\text{(109)}$$

The data obtained hitherto [255] are consistent with an internal anionic displacement of halogen (Eq. (110)).

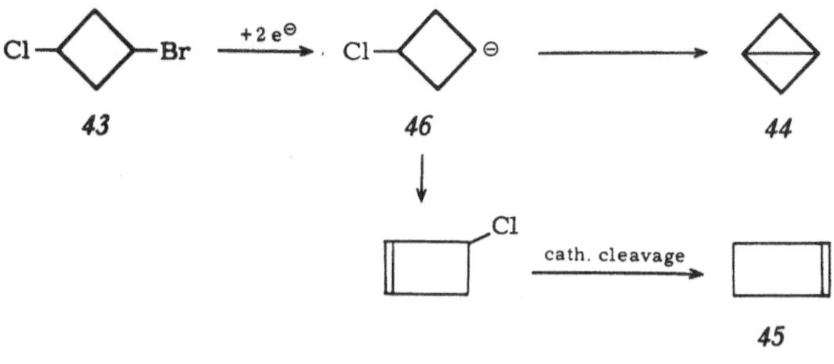

$$\text{(110)}$$

43 exhibits a polarographic wave at -1,8V (*v*.S.C.E.) corresponding to the reduction of the C-Br bond. Cpe at -2,0V requires two electrons per mole of *43* and yields 60% bicyclobutane (*44*), 20% cyclobutene (*45*), and 10% cyclobutane. *45* arises by E2-elimination from *43* with the intermediate anion *46* acting as the base and subsequent cathodic cleavage of hydrogen. With 1,3 dibromo-1,3-dimethyl-cyclobutane the proportion of the elimination product — analogous to *45* — increases with increasing concentration of the dibromide, as expected for an E 2-process. A strict distinction between a stepwise and a concerted anionic mechanism is not possible from as yet available data. A stepwise mechanism with an anion as intermediate is indicated by the concentration-dependent formation of the elimination product and similar reduction potentials for the *cis* - and *trans* isomers of 1,3-dibromo-1,3-dimethylcyclobutane. On the other hand, those dibromides which give cyclic products have a more positive reduction potential than those which give open chain hydrocarbons, which indicates a concerted mechanism.

Intramolecular substitution of H_2O^+ occurs in the reductive formation of tocopherol *(48)* from the protonated quinone *47* [256].

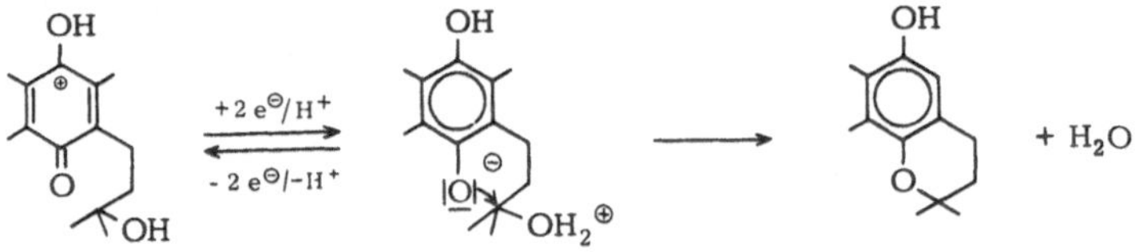

47   48

The cathodic replacement of substituents by hydrogen, a very useful preparative procedure of wide scope, is treated in Sect. 13.2.

10. Electrochemical Addition

10.1. Anodic Addition

The term anodic addition designates anode processes in which reagents, *e.g.*, nucleophiles Nu⁻, are oxidatively added to double bonds:

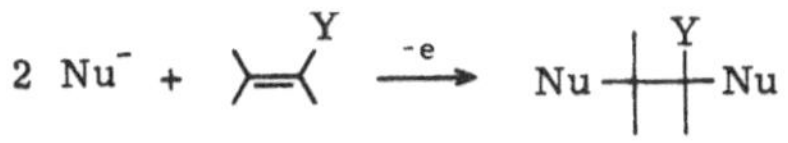

$$2\ \text{Nu}^- + \ \diagdown\!\!=\!\!\diagup^Y \ \xrightarrow{\ -e\ } \ \text{Nu}\!-\!\!\!\overset{\overset{\displaystyle Y}{|}}{\underset{|}{}}\!\!\!-\text{Nu}$$

Mechanistically three modes of addition can be distinguished:

a) *Addition of anodically generated radicals to double bonds (Eq. (111)):*

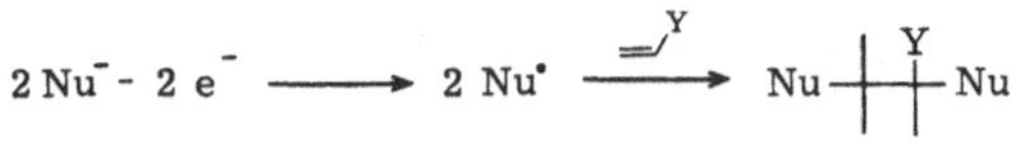

$$2\,\text{Nu}^- - 2\ e^- \ \longrightarrow \ 2\ \text{Nu}^{\bullet} \ \xrightarrow{\ =\!\!\diagup^Y\ } \ \text{Nu}\!-\!\!\!\overset{\overset{\displaystyle Y}{|}}{\underset{|}{}}\!\!\!-\text{Nu} \tag{111}$$

b) *Addition of anodically generated electrophiles to double bonds (Eq. (112)):*

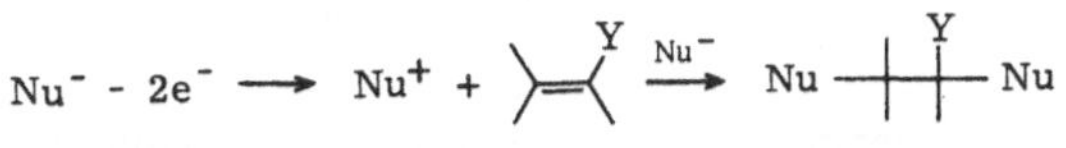

$$\text{Nu}^- - 2e^- \ \longrightarrow \ \text{Nu}^+ + \ \diagdown\!\!=\!\!\diagup^Y \ \xrightarrow{\ \text{Nu}^-\ } \ \text{Nu}\!-\!\!\!\overset{\overset{\displaystyle Y}{|}}{\underset{|}{}}\!\!\!-\text{Nu} \tag{112}$$

c) *Anodic additions via radical cations as intermediates (Eq. (113)):*

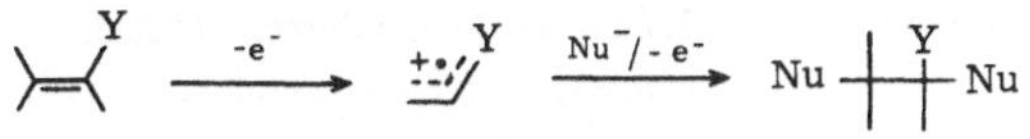

$$\diagdown\!\!=\!\!\diagup^Y \ \xrightarrow{\ -e^-\ } \ \overset{+\bullet}{\diagup}\!\!\diagup^Y \ \xrightarrow{\ \text{Nu}^-/-e^-\ } \ \text{Nu}\!-\!\!\!\overset{\overset{\displaystyle Y}{|}}{\underset{|}{}}\!\!\!-\text{Nu} \tag{113}$$

Anodic addition is a valuable preparative method since nucleophiles can be oxidatively added to unactivated double bonds. The manifold of synthetic subunits in the class of anions is thus no longer limited to use in nucleophilic reactions, but can be equally well applied in radical or electrophilic additions, whereby the choice of reagents for these reactions is considerably extended.

Electrochemical Addition

a) Addition of Anodically Generated Radicals to Double Bonds

The Kolbe electrolysis is a convenient source for alkyl radicals R· (Eq. (94)). If conducted in the presence of olefins, *e.g.*, butadiene, isoprene, cyclohexadiene, or styrene, the intermediate radicals R· add to the double bond to form adducts *49*, which dimerize to additive dimers *50* or couple with R· to disubstituted monomers *51* (Eq. (114)) [257-260].

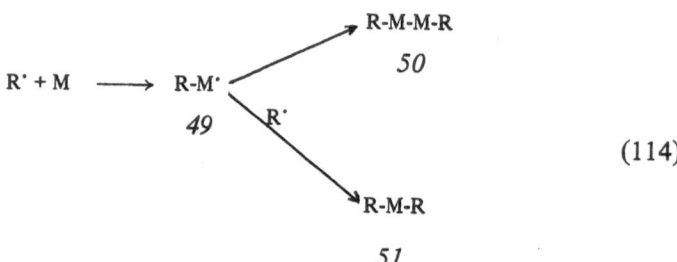

$$R\cdot + M \longrightarrow R\text{-}M\cdot$$

(114)

Some synthetic applications of this reaction are summarized in Table 7.

Table 7. *Addition of radicals, generated by electrochemical decarboxylation, to olefins*

Carboxylic acid	Olefin	Product	Ref.
Potassium ethyl oxalate (MeOH)	Butadiene	Diethyl 3,7-decadiene-1,10-dicarboxylate (40%)	[257]
Potassium acetate (acetic acid, MeOH)	Butadiene	*trans*-3-Hexene (11-26%), 1-pentene 3-methyl-1-pentene, C_{10}-dienes (12-58%), 3,7-decadiene, 3-ethyl-1,5-octadiene	[257,258]
Trifluoroacetate, trifluoroacetic acid, MeOH	Butadiene	1,1,1,10,10,10-Hexafluoro-3,7-decadiene, other products	[257]
Potassium acetate, acetic acid, MeOH	Isoprene	*cis*- 3-Methyl-3-hexene and other products	[258]
Potassium acetate, acetic acid, MeOH	Cyclohexadiene	Methylcyclohexane, *cis*- and *trans*-1,2-dimethylcyclohexene (42%), *trans*-1,4-dimethylcyclohexene (14%)	[259]
$Cl(CH_2)_4CO_2^-$	Butadiene	α, ω-Dichlorododecenes and hexadecadienes	[260]
Acetate, acetic acid	Styrene	3,4-Diphenylhexane and polymer	[262]
Methyl adipate, MeOH	Butadiene	Dimethyl 6,10-hexadecadiene-1,16-dicarboxylate (49%), dimethyl 6-dodecene-1,12-dicarboxylate (67,5%)	[263]

78

Additive dimerization is a useful synthetic method since four subunits, two radicals and two olefins, are joined in one step and in a specific way to head-to-head dimers *50a*. To obtain usable yields, however, precisely controlled reaction conditions have to be maintained.

The concentration of primary adduct *49* has to be high enough to favor dimerization (path d, Eq. (115)) and suppress the competing reactions, mixed

$$
\begin{array}{ccc}
\text{R-R} & & \text{R}\overset{Y}{\underset{}{\rule[-0.5em]{0pt}{1.2em}|}}\text{R} \\
& & 51a \\
\uparrow \text{a)} & \text{R}\cdot\text{c)} \nearrow & \\
\text{R}^{\bullet} \xrightarrow[\text{b)}]{\underset{}{=}\!\!\text{Y}} \text{R}\overset{Y}{\underset{}{|}}{\cdot} \xrightarrow{\text{d)}} \text{R}\overset{YY}{\underset{}{||}}\text{R} & & \\
52 \qquad\qquad\quad 49 \qquad\qquad 50a & & \\
& \underset{=}{\searrow}\,\text{Y} \searrow \text{e)} & \\
& & \text{R-}(\overset{Y}{\underset{}{|}})_{\overline{n}}\text{R}
\end{array}
\qquad (115)
$$

coupling (c) and polymerization(e) [261,262]. To achieve high concentration of *49* the concentration of *52* can only be raised to a certain optimal value. Otherwise coupling of R· (path a) will predominate over addition (path b), especially with olefins of low reactivity. In conventional radical chain additions [264a] (Eq. (116))

$$
\text{R}^{\bullet} + \underset{=}{\nearrow}\!\text{Y} \longrightarrow \text{R}\overset{Y}{\underset{}{|}}{\cdot}
$$

$$
\text{R}\overset{Y}{\underset{}{|}}{\cdot} + \text{RH} \longrightarrow \text{R}\overset{Y}{\underset{}{|}}\text{H} + \text{R}^{\bullet}
\qquad (116)
$$

radical concentrations are normally too low for additive dimer formation. As side products only telomers or polymers arise. If the radicals however are electrochemically generated the radical concentration in the reaction layer at the electrode-electrolyte interface is much higher than in homogenous radical additions. Thus radical concentrations favorable for additive dimerization are achieved and these can be further optimized by regulating the radical concentration *via* the current density at will. Therefore electrochemical radical addition is a unique way to prepare additive dimers. Only the metal ion initiated decomposition of peroxides in the presence of olefins [265a] gives similar results, but this method is more limited in suitable radical precursors and mostly restricted to water as solvent.

Anodic 1e-oxidation of organic and inorganic anions or organometallics is a preparatively simple route to most variably substituted radicals for addition to olefins. Schäfer *et al.* [264-268] have initiated a systematic study on the anodic addition of anions to olefins. Organic anions R⁻, *e.g.*, sodium salts of 1,3-dicarbonyl compounds, nitriles or aliphatic nitro compounds, sodium azide and Grignard reagents have been anodically added to olefins, *e.g.*, styrene, α-methylstyrene, butadiene, cyclohexene, cyclooctene, vinyl ethyl ether. The products obtained are best rationalized by the following scheme:

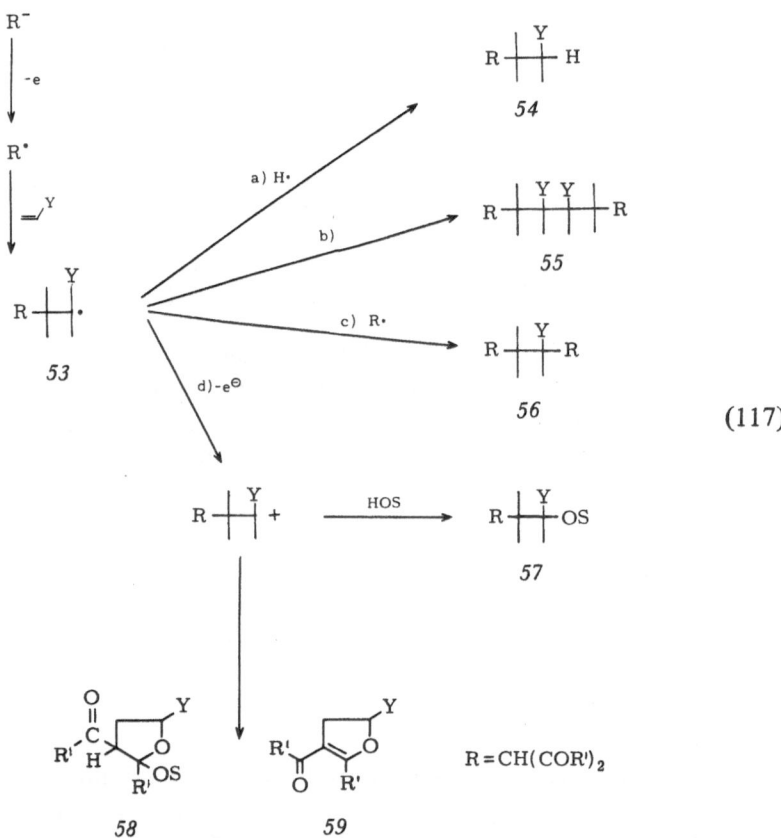

$$(117)$$

The primary adduct *53* (Eq. (117)) of the anodically generated radical R· undergoes a series of follow-up reactions: a) hydrogen abstraction to *54*, b) dimerization to additive dimers *55*,   c) coupling with R· to 1,2-disubstituted monomers *56*,   d) 1e-oxidation to a carbonium ion that either solvolyzes to *57* or, when 1,3-dicarbonyl compounds are added cyclizes intramolecularly to tetra (*58*) - or dihydrofuran derivatives (*59*). Product control is possible in some cases by suitable choice of the anode potential. With a high anode potential,

$>+ 0,8$ to $+ 1,2$ V *vs.* Ag/AgC1, the products *57, 58* and *59* are favoured for Y = Phenyl and formation of *55* is diminished, while at lower anode potentials, $<+ 0,5$ V, *55* predominates [265,268]. Synthetic applications of these reactions are, *e.g.*, a) with 1,3-dicarbonyl compounds: a combination of anodic addition with subsequent or preceding modification of the reagent by acylation, alkylation, keto or acid cleavage; b) with aliphatic nitro-compounds: subsequent reduction of the nitro group to an amino group or conversion to an acyl compound in a Nef-reaction; c) with Grignard reagents: one step construction of complex hydrocarbons from four subunits; d) with sodium azide: synthesis of 1,2- or 1,4-diazides for 1,3 dipolar addition, nitrene generation, or subsequent reduction to amines.

Table 8 illustrates the scope of anodic addition of anions to olefins:

Table 8. *Anodic addition of anions to olefins*

Anion	Olefin	Product	Ref.
1,3-Dicarbonyl compounds:			
Dimethyl sodio-malonate, methanol	Vinyl ethyl ether	Dimethyl 3-ethoxy-3-methoxypropane-1,1-dicarboxylate (37%)	264)
Dimethyl sodio-malonate, methanol	Styrene	Methyl 2,2-dimethoxy-5-phenyl-tetrahydrofuran-3-carboxylate (43%), tetramethyl 3,4-diphenylhexane-1,-1,6,6-tetracarboxylate (15%)	264)
Sodio-acetylacetonate, methanol	Vinyl ethyl ether	4-Acetyl-2-ethoxy-5-methyl-2,3-dihydrofurane (36%)	264)
Sodio-acetylacetonate, methanol	Butadiene	*trans, trans*-Tetradeca-5,9-diene-2,13-dione (40%)	264)
Nitriles:			
Methyl sodio-cyano-acetate, methanol	Styrene	Methyl 1-phenyl-1-methoxy-3-cyanopropane-3-carboxylate (18%)	265)
Methyl sodio-cyano-acetate, methanol	Vinyl ethyl ether	Methyl 1-ethoxy-1-methoxy-3-cyanopropane-3-carboxylate (19%)	269)

Table 8 (continued)

Ar ion	Olefin	Product	Ref.
Aliphatic nitrocompounds:			
Sodio-2-nitropropane, methanol	Styrene (cpe $<$0,5V Ag/AgCl)	2,7-Dimethyl-2,7-dinitro-4,5-diphenyl-octane (15%), 2,3-dimethyl-2,3-di-nitrobutane (35%)	265,268)
Sodio-2-nitropropane, methanol	Styrene (cpe $>$0,8V Ag/AgCl)	1-Methoxy-1-phenyl-3-nitro-3-methylbutane (45%)	265,268)
Sodio-2-nitropropane, methanol	Vinyl ethyl ether	3-Nitro-3-methyl-butyraldehyde ethyl methyl acetal (15%), 2,3-dinitro-2,3-dimethylbutane (45%)	265)
Grignard reagents:			
Butylmagnesium bromide, ether	Styrene	6,7-Diphenyldodecane (29%)	267)
t-Butylmagnesium-bromide	Styrene	4,5-Diphenyl-2,2,7,7-tetramethyloctane (14%)	267)
Butylmagnesium-bromide, ether	Butadiene	6,7-Divinyldodecane (3%), 6-vinyl-8-tetradecene (15%), 6,10-hexadecadiene (15%), 6-dodecene (7%)	267)
Sodium azide:			
Sodium azide, acetic acid	Styrene	1,4-Diazido-2,3-diphenylbutane (57%)	266)
Sodium azide, acetic acid	α-Methyl-styrene	1,4-Diazido-2,3-diphenyl-2,3-dimethyl-butane (45%)	266)
Sodium azide, acetic acid	Cyclooctene	1,2-Diazidocyclooctane (24%), azidocyc-looctane (15%), 3-azido-1-cyclooctene (17%)	266)

b) Addition of Anodically Generated Electrophiles to Double Bonds (see also Sect. 15.4)

This mode of anodic addition is so far limited to halogens. Cpe of cyclohexene in acetonitrile - $Et_4N^+Cl^-$ at the oxidation potential of the chloride ion yields 80% 1-chloro-2-acetamidocyclohexane[270], whose formation is rationalized by electrophilic addition of anodically generated chlorine to the double bond and subsequent Ritter reaction of the cationic intermediate (Eq. (118)). The same product is obtained by adding chlorine to a solution of cyclohexene in aceto-nitrile.

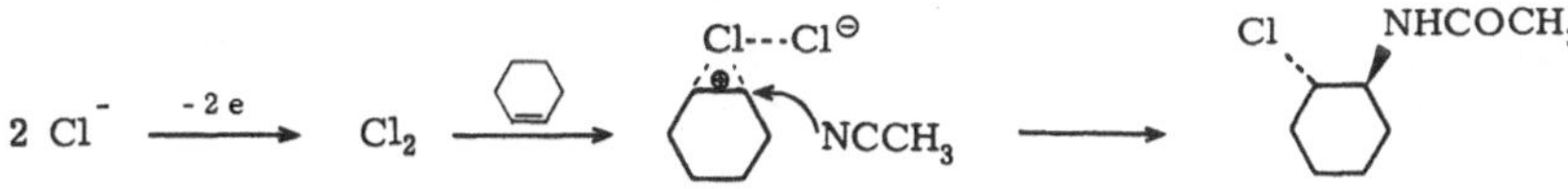

$$2 \ Cl^{-} \xrightarrow{-2e} Cl_2 \longrightarrow \square \longrightarrow \overset{Cl\cdots Cl^{\ominus}}{\underset{NCCH_3}{\square}} \longrightarrow \overset{Cl \quad NHCOCH_3}{\square} \qquad (118)$$

Analogous bifunctionalizations of cyclohexene, 1-phenyl-2-butene and cyclo-pentene to 1-halo-2-acetamido (acetoxy) adducts are reported by Weinberg and Hoffmann [270a].

A similar sequence probably accounts for the formation of 50% (40%) 1-chlo-ro- (1-bromo-) -2-methoxy-2-phenylpropionic acid by electrolysis of cinnamic acid in CH_3OH/NH_4Cl (NH_4Br) as SSE [271]. Electrolysis of *trans*-stilbene in me-thanol/NH_4Br produces *erythro*-stilbene dibromide and 1-methoxy-2-bromo-1,2-diphenylethane [104] in fair yields.

c) Anodic Addition *via* Radical Cations as Intermediates

This mode of anodic addition involves oxidation of the substrate RH, an olefin or aromatic compound, to a radical cation, in contrast to the oxidations of the rea-gent Nu⁻, depicted in the preceding sections a) and b). The adduct is formed in a EC_NEC_N- sequence (Eq. (119)):

$$RH \xrightarrow[E]{-e} RH^{+\cdot} \xrightarrow[C_N]{Nu^-} RHNu^{\cdot} \xrightarrow[E]{-e} RHNu^{+} \xrightarrow[C_N]{Nu^-} RHNuNu \qquad (119)$$

In this way furanes form 2,5-dialkoxy-2,5-dihydrofuranes in high yields on electrolysis in R'OH/NH_4Br (Eq. (120)):

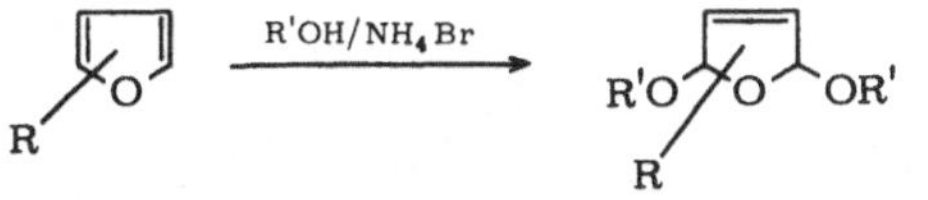

$$R:H, \ R':CH_3 \ 73\% \ ^{[272]}$$
$$R:CH_2OAc, \ R':CH_3 \ 87\% \ ^{[273]}$$
$$R:CO_2Et, \ R':Et \ 62\% \ ^{[274]}$$

(120)

With suitable substituents intramolecular alkoxylation to spiroketals can be achieved (Eq. (121)) [275]. For details on alkoxylation of furans the reader is re-ferred to extensive Tables in Ref. [10].

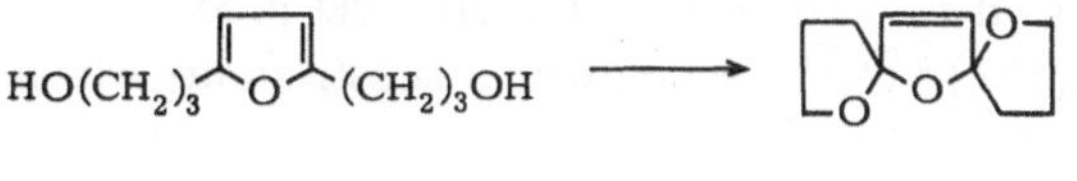

$$HO(CH_2)_3 \overset{\square}{\underset{O}{\diagup}} (CH_2)_3OH \longrightarrow \qquad \qquad (121)$$

Ross *et al.* [105] obtained a 1:1 mixture of *cis-* and *trans*-2,5-dimethyl-2,5-di-methoxy-2,5-dihydrofurane (*60*) from 2,5-dimethylfurane both by electrochemical and chemical oxidation. Besides *60* minor amounts of 5,5'-bis (2-methoxy-2-methyl-2,5 -dihydrofurane) (*61*), the dimer of the intermediate radical cation, were isolated. 2-Cyano-5-methoxy-2,5-dimethyl-2,5-dihydrofurane was obtained on electrolysis of 2,5-dimethylfurane in methanol-NaCN [276].

Analogous alkoxy-adducts are obtained with alkoxylated benzenes or pyridines (Table 9) *89*.

Table 9. *Anodic additions to aromatics*

Aromatic	Product	Ref.
Hydroquinone dimethyl ether	3,3,6,6-Tetramethoxy-1,4-cyclohexadiene (88%)	277)
1,2,4-Trimethoxybenzene	2,3,3,6,6-Pentamethoxy-1,4-cyclohexadiene (89%)	277)
Hydroquinone bis-hydroxyethylether	 (60 - 70%)	278)
2,5-Dimethoxypyridine	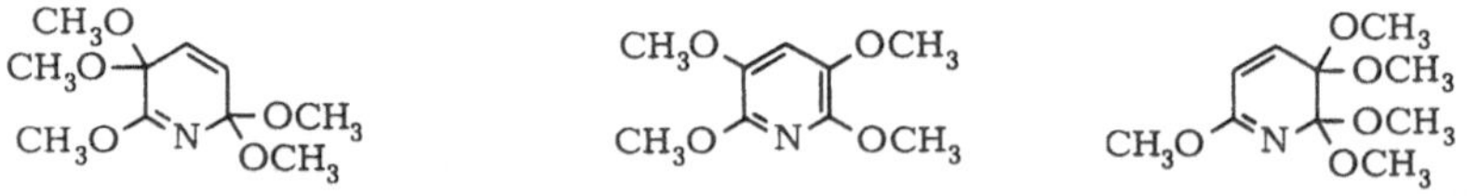	279)

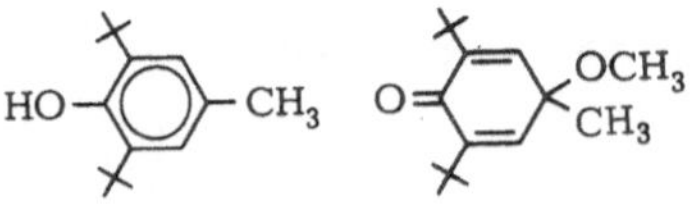

(CH$_3$CN, NaClO$_4$, Et$_4$NOH)

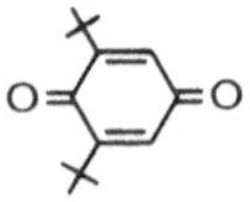

Electrolysis of substituted phenols in acetic acid produces quinolacetates in good yields (Eq. (122)) [281]. Phenols with *p*-substitutents, which bear a carboxylgroup in β-position, form spirolactones by intramolecular cyclization (Eq. (123) [282]).

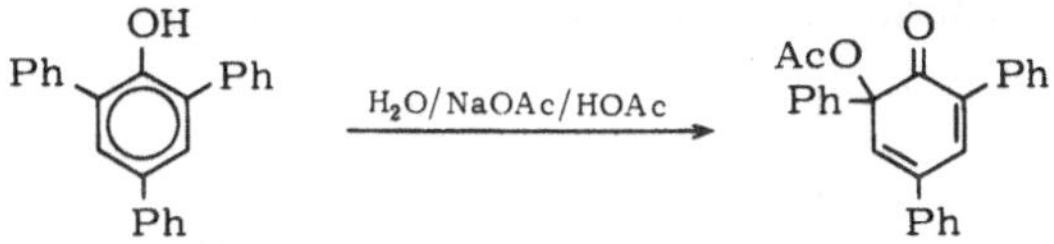

$$(122)$$

$$(123)$$

Pyridine may react as a nucleophile with radical cations to pyridinium salts, as in the electrolysis of anthracene in acetonitrile-pyridine which yields a bis-pyridinium-adduct [283]. With tris (p-methoxyphenyl)ethylene a tripositive tripyridinium cation is formed (Eq. (124)) [284].

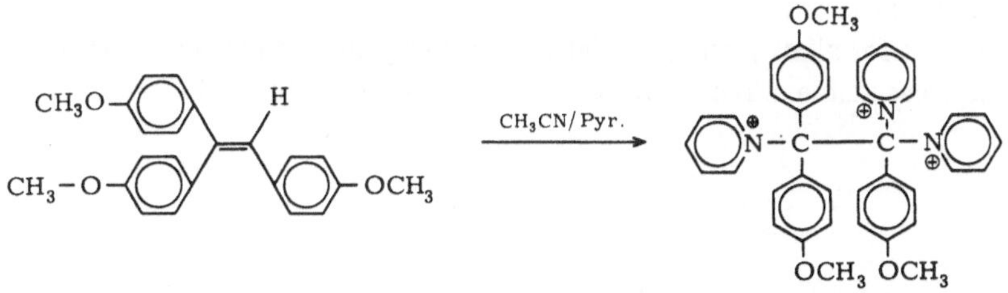

$$(124)$$

Acyloxylation of aryl olefins probably involves radical cations as intermediates. Acetoxylation of *trans*-stilbene in anhydrous acetic acid/sodium acetate yields mainly *meso*-diacetate, while in moist acetic acid mainly *threo*-2-acetoxy-1,2-di-phenylethanol is formed [100]. Anodic oxidation of *trans*- and *cis*-stilbene in acetonitrile/benzoic acid produces with both olefins the same mixtures of *meso*-hydrobenzoin diacetate (*62*) and *threo*-2-benzoyloxy-1,2-diphenylethanol (*63*) [101]. Product formation is best rationalized by a EC$_N$E-sequence leading to the energetically most favorable acyloxonium ion (*64*) (Eq. (125)):

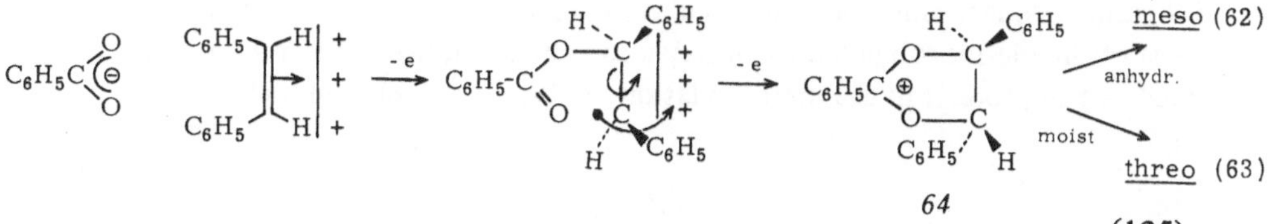

$$(125)$$

1,1-Diphenylethylene forms 15% 1,2-diacetoxy-1,1-diphenylethane and 59% 2-acetoxy-1,1-diphenylethanol-1 [100]. Cpe of cyclooctatetraene at +1,5 V in acetic acid at a platinum electrode yields the products *65*, *66* (7,7%) and *67–69*

85

(35%, ratio: 3:5:3). With a carbon anode no methyl substituted products are found, but 61% diacetates *67–69* [102]. The 8-acetoxyhomotropylium cation might be an intermediate in this acetoxylation (see Eq. (33)).

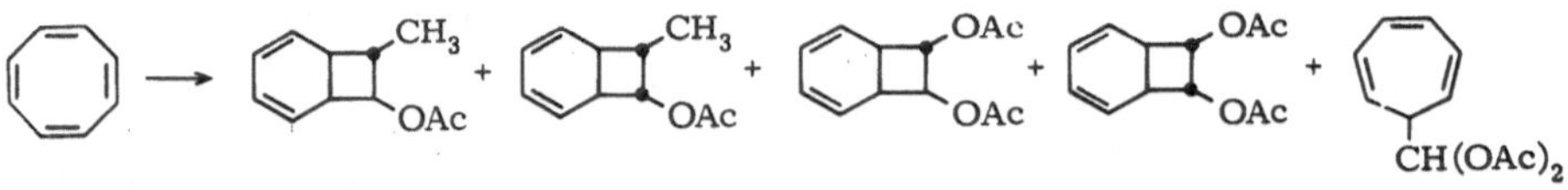

65	*66*	*67*	*68*	*69*

1,3-Cyclohexadiene produces 1,2- and 1,4-diacetoxycyclohexene [285], cyclohexene forms 3-acetoxycyclohexene [285], and 2-methyl-2-butene yields 2-methyl-3-acetoxy-1-butene [286] on electrolysis in acetic acid/sodiumacetate as SSE, presumably *via* radical cations.

Although also for alkoxylation of olefins a pathway *via* radical cations (Eq. (119)) is very reasonable, a radical route (Eq. (126)) cannot be excluded. This

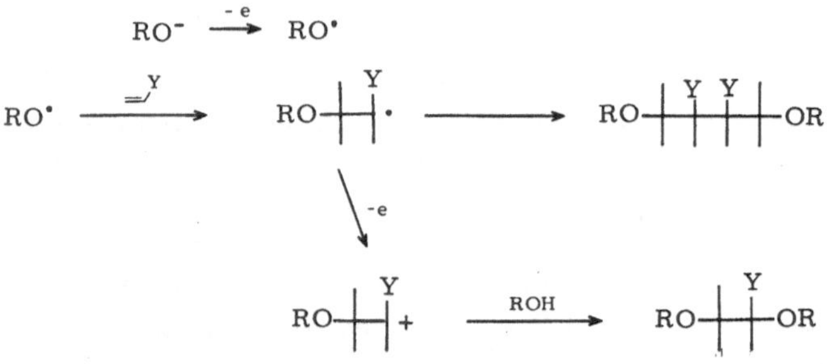

$$(126)$$

is because primary and secondary alcohols have a lower discharge potential than most of the olefins subjected to alkoxylation and an additive dimer *71* was isolated as side product in the methoxylation of styrene at a platinum electrode [287].

$$CH_3O\text{-}CH_2\text{-}\underset{\underset{C_6H_5}{|}}{CH}\text{-}\overset{\overset{C_6H_5}{|}}{CH}\text{-}CH_2\text{-}OCH_3$$

71

Table 10 gives some examples of olefin alkoxylations:

Table 10. *Anodic alkoxylation of olefins*

Olefin	Product	Ref.
trans-Stilbene	Hydrobenzoin dimethyl ether (*meso* (26%) *d*, *1* (18%))	288)
cis-Stilbene	Hydrobenzoin dimethyl ether (*meso* (20%) *d*, *1* (44%))	288)
Styrene	Styrene glycol dimethyl ether, *meso*-1,4-dimethoxy-2,3-diphenylbutane	287)
Diphenylethylene	1,1-Diphenylethylene glycol dimethyl ether	287)
Norbornene	*exo, syn*-2,7-Dimethoxy-bicyclo [2.2.1] heptane *exo*- 2-methoxybicyclo [2.2.1] heptane	104)
1,3-Cyclohexadiene	1,2-Dimethoxy-3-cyclohexene,1,4-dimethoxy-2-cyclohexene	285)
Cyclohexene	3-Methoxycyclohexene, cyclopentanecarboxaldehyde dimethylacetal	285)

Carbomethoxylation of olefins to α, β-unsaturated esters or β-methoxy-esters (Eq. (127) can be achieved if the electrolysis in $CH_3OH/NaOCH_3$ at platinum electrodes is conducted under CO pressure. A platinum carbonyl-complex $Pt_x (CO)_y$ is assumed to be the reactive intermediate in this synthetically interesting reaction [289].

$$PhCR=CH_2 + CO + CH_3O^- \longrightarrow \begin{array}{c} \overset{R}{\underset{|}{PhC=CHCOOCH_3}} \\ \\ \underset{|}{PhCR-CH_2COOCH_3} \\ OCH_3 \end{array} \qquad (127)$$

R: H 14%, CH_3 14,8%, C_6H_5 30%
R: H 3,6%

Alkoxylation of enol ethers to α-alkoxy acetals [35] certainly involves radical cations as intermediates as tail-to-tail dimers, which exclude the radical path, (Eq. (126)) are formed as side [35] or main products [36,268].

10.2. Cathodic Addition

In cathodic addition reactions solvated electrons, radical anions or anions are generated at the cathode and added to activated or unactivated double or triple bonds. This broad spectrum of reactions is partially treated a) in section 8.2 when group conversion generates a reactive intermediate which undergoes addition reactions [155a,b], and b) in section 12.2 when olefins are coupled *via* addition of a cathodically generated radical anion to an activated double bond.

The solvated electron [290] is a very useful reagent for the reduction of double bonds. Benzene, toluene or *p*-xylene selectively form 1,4-dihydro products (72 in Eq. (128)) with less than 5% 1,2-dihydro product, when they are electrolyzed in glyme (tetrahydrofurane, diglyme) — 10% water and tetrabutyl-ammonium perchlorate as SSE [291]. Naphthalene is reduced at the mercury cathode in acetonitrile/25% water/tetraethylammonium tosylate in 80% current efficiency almost exclusively to 1,4-dihydronaphthalene [292]. The high selectivity for 1,4-dihydro products indicates that the unconjugated dienes do not isomerize to conjugated dienes, the 1,2-dihydro products, under the reaction conditions in contrast to the less selective reduction of aromatics by alkali metal in liquid ammonia. The solvated electron is assumed to be the reactive reagent. Bound to the tetraalkylammonium cation it is transferred to the aromatic compound yielding a radical anion, which by subsequent protonation-reduction-protonation forms the 1,4-dihydro product (72) (Eq. (128)):

$$R_4N^+ \xrightarrow{+e} R_4N^+ \text{- - -} e^{\ominus}_{solv.}$$

$$R_4N^+ \text{- - -} e^{\ominus}_{solv.} + \text{⬡} \longrightarrow \text{⬡}^{\ominus} \xrightarrow{H^+} \text{⬡}^{\cdot} \xrightarrow[H^+]{+e} \text{⬡} \quad (128)$$

72

This mechanism is reasonable as a) reduction of benzene occurs at a cathode potential of -2,5 V *vs.* S.C.E., roughly corresponding to the standard potential of the hydrated electron [293], while the potential for the direct electron transfer to benzene is more negative (~-3,0 V) and b) *"in situ"* electrolysis in the ESR cavity produces at -100 °C the characteristic singlet of the solvated electron [293a], which changes to the septett of the benzene radical anion, when benzene is added to the solution.

A mixture of 67% ethanol and 33% hexamethylphosphortriamide (HMPTA), 0.3 M in lithium chloride, is an excellent SSE for the generation of solvated electrons [294]. In this solvent tetraline is hydrogenated to decaline, octaline and hexaline (Eq. (129)).

$$\text{(129)}$$

Benzene is reduced in 95% current yield to a mixture of 23% cyclohexadiene, 10% cyclohexene and 67% cyclohexane. HMPTA as a solvent additive seems to play a dual role. Firstly it is selectively adsorbed at the cathode surface, thereby preventing hydrogen evolution from the protic solvent. Thus it permits the attainment of a potential sufficiently cathodic for the generation of the solvated electron. It secondly stabilizes the solvated electron, thus suppressing its reaction with protic solvents (eq. (130)). With decreasing HMPTA concentration in the electrolyte the current efficiency for reduction decreases and hydrogen evolution dominates. In pure ethanol the current efficiency is less than 0,4%.

$$e_s^- + C_2H_5OH \longrightarrow 1/2\, H_2 + C_2H_5O^- \qquad (130)$$

The selectivity for di-, tetra- and hexahydro products can be controlled by properly adjusting current density, temperature and proton availability. At -20 $^\circ$C and 3,6 mA/cm^2 in 67% EtOH/33%HMPTA, benzene is converted in 93% yield to cyclohexadiene, while at 28 $^\circ$C and 91 mA/cm^2 20% cyclohexadiene, 8,5% cyclohexene and 71% cyclohexane are obtained. At high proton availability also highly sterically hindered olefins, *e.g.*, tetramethylethylene, are readily reduced. It generally holds that the higher the proton availability, the more complete is the hydrogenation. Even bituminous coal can be hydrogenated with an uptake of 53 hydrogens for 100 carbons. When in the reduction of naphthalene the solvent composition was slightly changed, a drastic change in product was achieved. In 30% HMPTA/70% MeOH(EtOH) 100% 1,4-dihydronaphthalene was obtained, while with 30% HMPTA/70% isopropanol as solvent 90% 1,2-dihydronaphthalene was obtained in 100% current yield [295]. Such product variations by solvent change have been attributed to modifications of the electric field in the electrical double layer [296].

Effective product control can be achieved by properly adjusting the basicity of the electrolyte [297]. Reduction of benzene with LiCl/methylamine as SSE in an undivided cell yields exclusively 1,4-cyclohexadiene, while in a divided cell cyclohexene is formed (Eq. (131)). This difference is attributed to different basicities of the electrolytes. In the divided cell the lithium alkylamide formed isomerizes 1,4-cyclohexadiene to 1,3-cyclohexadiene, which is further reduced to cyclohexene.

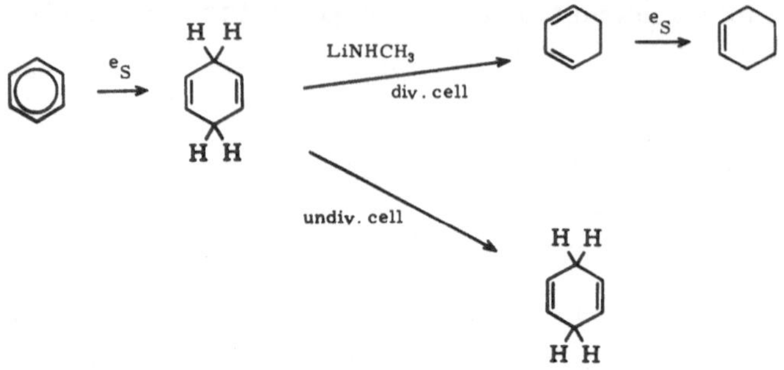

$$(131)$$

In the undivided cell the base is neutralized by anodically generated methyl-ammonium chloride and isomerization is thus prevented. 1,4-Cyclohexadiene itself is not reduced under the conditions employed.

In the presence of ethanol to neutralize traces of Li-alkylamides, which would cause isomerization, a selective reduction of an aromatic ring in the presence of an isolated double bond is possible [299]. Examples are:

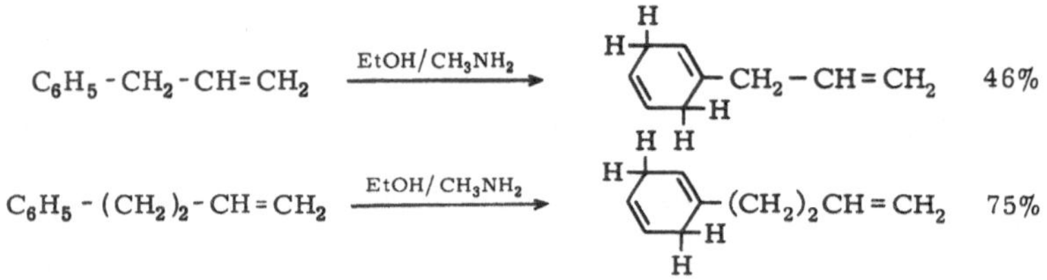

$$(132)$$

2- and 3-octyne and 5-decyne are converted in 50–80% yield to *trans*-olefins on electrolysis in LiCl/methylamine [298]. From conjugated alkynes the saturated hydrocarbons were obtained, *e.g.*, phenylacetylene gave ethylbenzene, 1-phenyl-1-butyne butylbenzene. When the aryl group is further removed from the triple bond, *e.g.*, in 4-phenyl-1-butyne or 5-phenyl-2-butyne, *trans*-olefins are the major products. The aromatic steroid *73* has been reduced in ethylenediamine/LiCl as SSE to the dihydro derivative *74* [300].

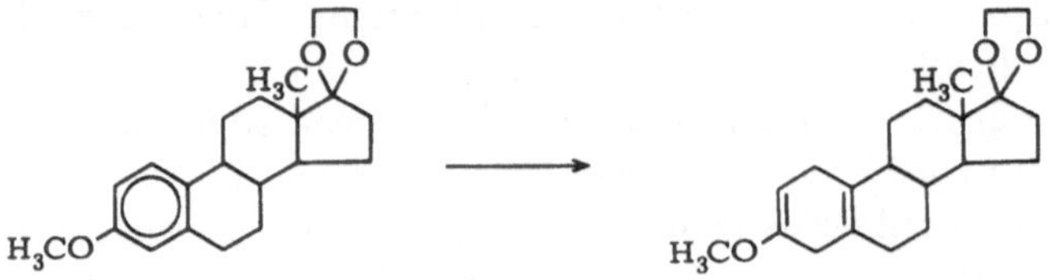

73   74

1°-, 2°- or 3°-amides are cleanly converted to the corresponding alcohols in

65–75% yield [299] (Eq. (133)). This reduction is clearly superior to lithium reduction, whose application is limited to 3° amides.

$$\underset{\substack{\| \\ O}}{RC\text{-}NH_2} \xrightarrow{\text{LiCl/MeNH}_2} RCH_2OH \tag{133}$$

Hydrogenation of double bonds can also be performed at the mercury cathode using methanol/tetramethylammonium salts as SSE [301]. The solvated electron, stabilized by $(CH_3)_4N^+$ and transferred in the electrical double layer to the double bond, is assumed to be the reductant. As in $CH_3NH_2/LiCl$ conjugated double bonds may be reduced, while isolated ones remain untouched. Terminal triple bonds are hydrogenated to double bonds.

The stereochemistry of cathodic hydrogenation is thus far not predictable or explainable. There is a preponderance for *trans* addition in the reduction of alkynes, while *cis* addition occurs predominantly with *cis*- and *trans*-stilbene and maleic acid derivatives. On the other hand, dimethylmaleic acid and dimethyl-fumaric acid are exclusively hydrogenated to the *trans* products, *meso*- and *dl*-dimethylsuccinic acid [302]. With a silver-palladium cathode in 5% aquous sulfuric acid triple bonds are reduced in a specific *cis* addition to double bonds [303].

Anions generated by cathodic reduction of carbonyl compounds, benzyl chloride or activated double bonds can be trapped by addition to CO_2. With benzophenone the carbonate of benzilic acid was obtained (Eq. (134)) [304].

$$\underset{\substack{\| \\ O}}{C_6H_5C\,C_6H_5} \xrightarrow{+2e/CO_2} \underset{\substack{| \\ OCO_2^-}}{\overset{CO_2^-}{\overset{|}{C_6H_5C}}}\text{---}C_6H_5 \tag{134}$$

The benzyl anion generated from benzyl chloride gave a low yield of phenylacetic acid [305]. The radical anion dimer of benzalacetone 75 reacts simultaneously by intramolecular addition to the carbonyl group and intermolecular addition to CO_2 to the cyclopentane derivative 76 [306] (Eq. (135)).

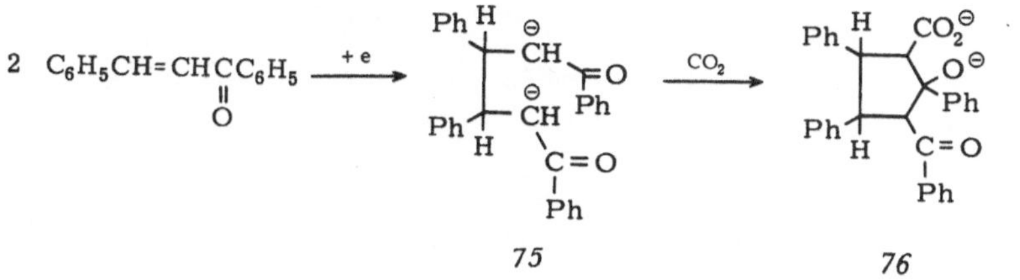

$$2\ \ \underset{\substack{\| \\ O}}{C_6H_5CH=CHCC_6H_5} \xrightarrow{+e} \quad \mathbf{75} \xrightarrow{CO_2} \quad \mathbf{76} \tag{135}$$

Electrochemical Addition

Electrolysis of benzalacetophenone in the presence of CO_2 gave α-phenylbenzoylpropionic acid [306]. Reduction of stilbene in the presence of CO_2 yielded 92% *meso*-2,3-diphenylsuccinic acid [307]. The reduction of anthracene at -2,3 V in the presence of SO_2 and O_2 yields 9,10-dihydroanthracene-9, 10-disulfonic acid [307a]. An interesting intramolecular addition was found in the reduction of the thiocyanato steroid *77* to the mercaptoimine *78* [308].

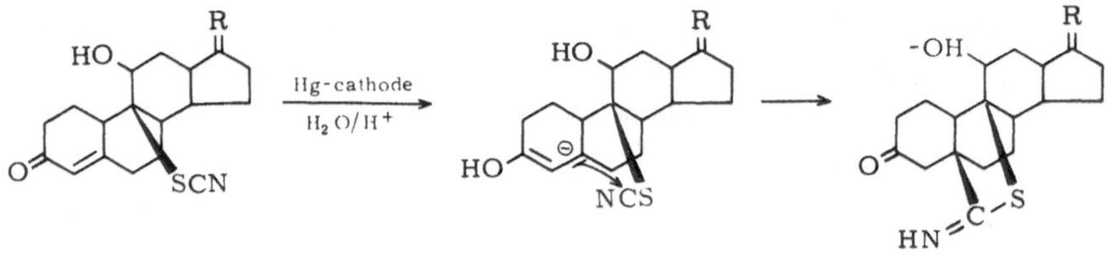

77　　　　　　　　　　　　　　　　　　　*78*

11. Electrochemical Elimination

11.1. Anodic Elimination

In anodic elimination reactions two substituents X, *e.g.*, hydrogen or CO_2, are oxidatively split off from the substrate to yield double or triple bonds (Eq. (136))

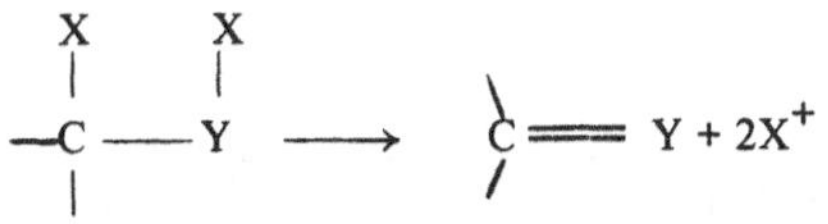

$$\underset{\substack{| \\ X: H, CO_2^- \\ Y: C, O}}{\overset{\substack{X \quad X \\ | \quad | }}{-C - Y}} \longrightarrow \quad \overset{\textstyle \diagdown}{\underset{\textstyle \diagup}{C}} = Y + 2X^+ \tag{136}$$

Anodic dehydrogenations, *e.g.*, oxidations of alcohols to ketones, have been treated in Sect. 8.1 and formation of olefins by anodic elimination of CO_2 and H^+ from carboxylic acids was covered in Sect. 9.1. Therefore this section is only concerned with anodic bisdecarboxylations of *vic*-dicarboxylic acids to olefins. This method gives usually good results when its chemical equivalent, the lead tetraacetate decarboxylation, fails. Combination of bisdecarboxylation with the Diels-Alder reaction or [2.2]-photosensitized cycloadditions provides useful synthetic sequences, since in this way the equivalent of acetylene can be introduced in cycloadditions.

Anodic bisdecarboxylation is normally conducted at a platinum anode in water-pyridine-triethylamine as electrolyte and with both carboxyl groups neutralized. Synthetic applications of this method are illustrated by the following examples (Eq. (139-145)).

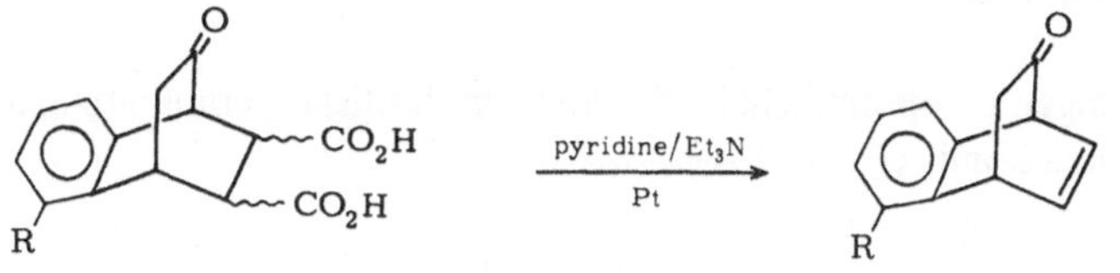

$$\tag{139}$$

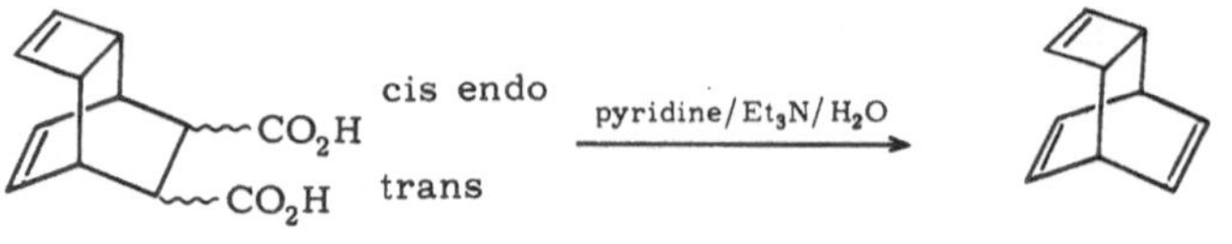

Electrochemical Elimination

$$cis \ endo \quad \xrightarrow{\text{pyridine/Et}_3\text{N/H}_2\text{O}} \quad 37\% \ [310]$$

trans $\quad 53\%$ $\qquad$ (140)

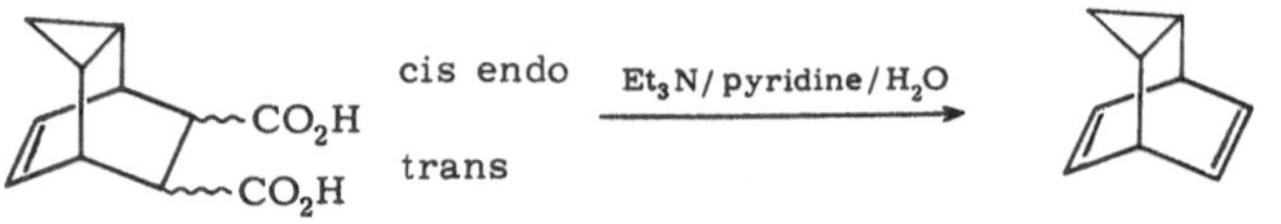

$$cis \ endo \quad \xrightarrow{\text{Et}_3\text{N/pyridine/H}_2\text{O}} \quad 36\% \ [310]$$

trans $\quad 45\%$ $\qquad$ (141)

$$35\% \ [311] \qquad (142)$$

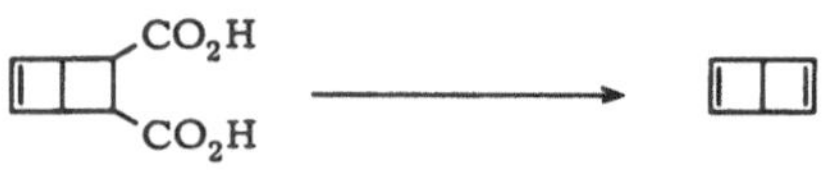

$$35\% \ [311] \qquad (143)$$

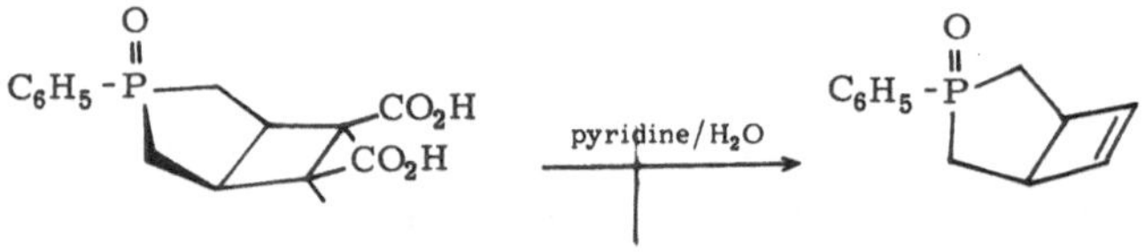

$$\xrightarrow{\text{pyridine/H}_2\text{O}} \quad 64\% \ [312] \qquad (144)$$

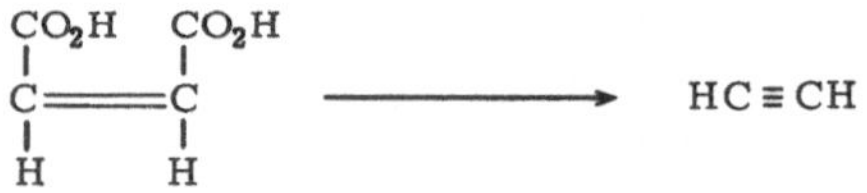

$$\longrightarrow \quad HC \equiv CH \qquad 5\% \ [313] \qquad (145)$$

A two step process involving an intermediate radical or carbonium ion has been suggested as mechanism because of the fact that both *meso-* and *d,1-*2,3-diphenylsuccinic acid yield 35-40% *trans*-stilbene [314]. A cationic intermediate however seems improbable for the bisdecarboxylations of aliphatic bicyclic carboxylic acids (Eq. (139-144)), since no products with rearranged carbon skeleton, as expected from carbonium ions, have been isolated.

11.2. Cathodic Elimination

From vicinal dihalogen compounds halogen can be cathodically eliminated to yield double or triple bonds (Eq. (146)).

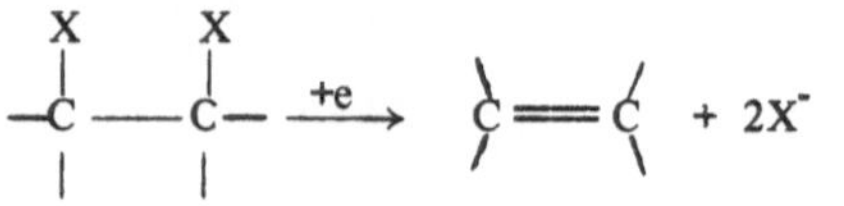

$$\underset{\underset{|}{X}}{\overset{\underset{|}{X}}{-C-C-}} \xrightarrow{\text{+e}} \ \ \overset{}{\underset{}{C}}=\overset{}{\underset{}{C}} \ + \ 2X^- \qquad (146)$$

With perchlorinated substrates first dechlorination to vinyl chlorides occurs and then cathodic cleavage of residual chlorine (Sec. 13.2) with subsequent hydrogenation of the double bond (Eq. (147)) [315]. Electrolysis of 1-iodo-2-bromoethane produces ethylene, whereas 1,1,1-trifluoro-2-bromochloroethane yields 70% 1,1-difluoro-2-chloroethylene [316]. Ethylene and butylene were ob-

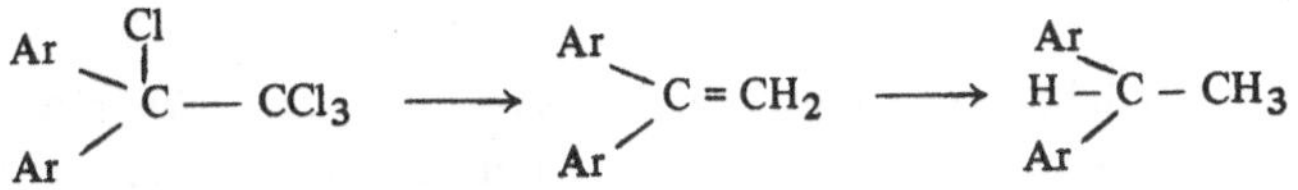

$$\underset{Ar}{\overset{Ar}{>}}\!\!\overset{Cl}{\underset{|}{C}}\!-CCl_3 \longrightarrow \underset{Ar}{\overset{Ar}{>}}C=CH_2 \longrightarrow \underset{Ar}{\overset{Ar}{>}}\!\!\overset{|}{\underset{|}{C}}\!-CH_3 \qquad (147)$$

tained from 1,2-dibromoethane or 2,3-dibromobutane respectively [317]. From *meso*-2,3-dibromosuccinic acid fumaric acid is formed exclusively by electrolysis at different pH-values. With the *d,l*-form at pH $<$ 0,4 and pH $>$ 6,9 fumaric acid is produced too whereas at pH 4 a maximum yield of 70% maleic acid is obtained [318]. This behaviour is attributed to a fixed conformation *79* in the succinic acid due to an internal hydrogen bond in the monoanion at pH 4 (see however reference [318a], regarding the hydrogen bond in *79*). Hexafluorobenzene has been prepared in 85 to 95% yield by reducing octafluoro-1,3- or 1,4-cyclohexadiene at a mercury cathode [319].

Similarly pentafluoro- and tetrafluorobenzenes were obtained from suitable fluorocyclohexadienes. *Vic*-cyano groups can be eliminated by solvated electrons [320].

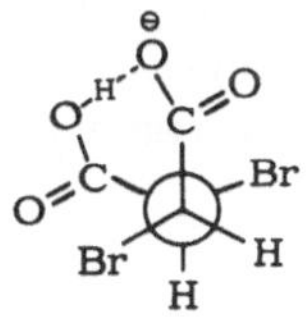

79

Electrolysis of 1,2-dibromofumaric and 1,2-dibromomaleic acid or the diesters produces a 90-100% yield of acetylenedicarboxylic acid or diester [302]. For *vic*-dihalides with no radical or anion stabilizing group in the α position an E_{2B} like elimination mechanism is strongly indicated, *i.e.,* 2e-transfer and double bond formation in a synchronous process (Eq. (148)).

$$\xrightarrow{+2e} \quad >\!\!=\!\!< \quad + \quad 2\,X^{\ominus} \qquad (148)$$

Vic -dihalides exhibit a single 2e polarographic wave, while nonvicinal ($n > 3$) dibromides normally are reduced in two separate 2e transfers. The reduction potential of the dihalides is more positive ($\triangle E \sim +0.4$ to $+0.7$ V) than that of the monohalides [316,321,322]. A plot of the half-wave potentials of 21 vicinal dibromides versus the dihedral angle, φ, between the two C-Br bonds exhibits maxima (most positive $E_{1/2}$) for $\varphi = 180°$ (*anti-periplanar* conformation) and $\varphi = 0°$ (*syn-periplanar* conformation) and a minimum (most negative $E_{1/2}$) for $\varphi = 90°$ [321].

By cathodic 1,6-dehalogenation *p*-xylylenes were prepared. When $\alpha,\alpha,\alpha,\alpha',\alpha',\alpha'$- - hexachloro-*p*-xylene was electrolyzed at low temperatures the tetrachloro-*p*-xylene could be isolated [323]. Cpe of α, α'-dibromo-*p*-xylene at the plateau of the first or second polarographic wave yields 5-10% [2.2] -paracyclophane and 90% poly-*p*-xylylene as products from the reactive unsubstituted *p*-xylylene (Eq. (149)) [322].

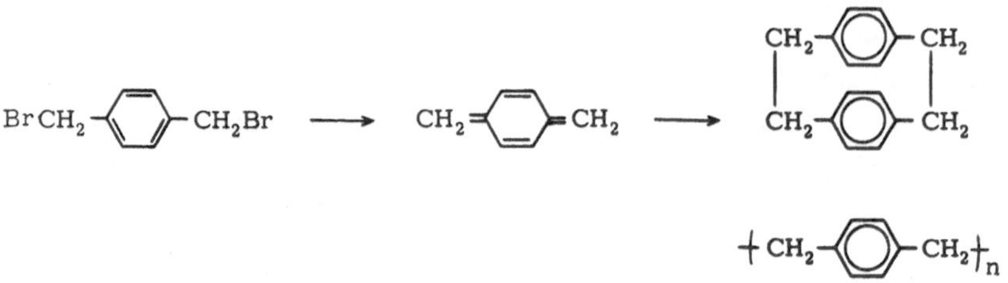

$$BrCH_2-\langle\bigcirc\rangle-CH_2Br \longrightarrow CH_2=\langle\bigcirc\rangle=CH_2 \longrightarrow$$

(149)

Debromination of 1,2-dibromobenzene gave benzyne in low yield [173]. Similarly dechlorination of carbon tetrachloride produced dichlorocarbene that could be trapped in low efficiency by tetramethylethylene (see also Sect. 15.5) [172]. The low yields of benzyne and carbene were partially attributed to their fast subsequent reduction.

12. Electrochemical Coupling

12.1. Anodic Coupling

Anodic coupling can be achieved in two ways. These are
a) *Coupling of anodically generated radicals (Eq. (150)):*

$$2\,R^- \xrightarrow{-e^-} 2\,R^{\cdot} \longrightarrow R\text{-}R \tag{150}$$

and
b) *Coupling of anodically generated radical cations (Eq. (151)):*

$$2\,RH \longrightarrow 2\,RH^{+\cdot} \longrightarrow {}^+HR\text{-}RH^+ \begin{array}{c} \xrightarrow{-H^+} R\text{-}R \\ \\ \xsearrow{Nu^-} HNuR\text{-}RHNu \end{array} \tag{151}$$

a) Coupling of Anodically Generated Radicals

The electrochemical reaction which is most intensively studied and has the widest synthetic applications is the Kolbe electrolysis of carboxylates (Eq. (152)) to give dimers (*80*):

$$2\,RCO_2^- \xrightarrow{-2e} 2\,CO_2 + 2\,R^{\cdot} \longrightarrow R\text{-}R \tag{152}$$
$$80$$

The discussion of this reaction can be limited to a short summary (see Table 11 for some representative examples) since there are exhaustive and critical reviews on both the mechanism [14,21,324] and the synthetic applications [14,325, 326] of the Kolbe electrolysis available.

The yield of Kolbe dimer (*80*) is determined by experimental and structural factors. The following experimental factors are important:

a) application of a high positive anode potential ($>$ +2,1 V v. S.C.E.) and high current densities ($>$ 0,25 A/cm^2);

b) in aqueous medium only smooth platinum or iridium as anode material yields dimers, while with carbon electrodes carbonium ion based products are predominant (see Sect. 10.1, Eq. (94)). In nonaqueous solvents the choice of the anode material is less critical though smooth platinum is preferred here too;

c) optimum yields in coupling product are obtained if the electrolyte is kept slightly acidic during the electrolysis which is achieved by neutralizing only a small amount of the carboxylic acid;

d) yields can be raised by using nonaqueous solvents, *e.g.*, methanol, acetonitrile, dimethylformamide or acetic acid, instead of water [327];

e) an increase in temperature decreases the yield of dimer;

f) foreign electrolytes, *e.g.*, F$^-$, SO$_4^{2-}$, ClO$_4^-$, Fe^{2+}, Cu^{2+}, Mn^{2+}, as additives inhibit or suppress the coupling process.

The following structural features influence the yield of coupling product:

a) Allyl, aryl, hydroxy, halogen, amino or acylamino groups as substituents in the α-position decrease or totally suppress the formation of coupling products and favor the carbonium path (Eq. (94)); for exceptions see Ref. [327a] and [328a];

b) Carboxamido [328], cyano [329] or carbalkoxy [330] groups in α-position favor the formation of coupling products;

c) substituents in γ to ω position usually exert negligible influence on the yield in coupling product.

The electrochemical mechanism of the Kolbe electrolysis, though very intensively studied, is still a matter of debate. For successful coupling a critical potential, around +2,1 V, has to be exceeded; otherwise oxygen evolution occurs and dimerization is totally suppressed. Conway *et al.* [21,331] attribute this behaviour to a coverage of the electrode surface with adsorbed acyloxy radicals both in the anhydrous and the aqueous case. These adsorbed radicals form a film on the anode surface which is a prerequisite for the Kolbe reaction. Fleischmann *et al.* [332] have studied the process using a repetitive potential pulse technique, and postulate formation of a platinum oxide layer as the rate and potential determining step for the Kolbe reaction. Eberson [198] invoked a concerted mechanism with simultaneous electron transfer and decarboxylation as potential determining step. On the other hand Skell [333] has postulated the formation of RCO$_2$· in the rate determining step as the nature of R in RCO$_2^-$ in competition experiments exerts no influence on the relative electrolysis rates. This interpretation has been criticized, however [334].

In the chemical reaction Conway *et al.* [21] assumes metal adsorbed radicals as reactive intermediates, while Eberson [324] advocates free radical intermediates. Though reactions *via* adsorbed radicals cannot be rigorously ruled out, such interactions must be very weak as the anodically generated radicals react similarly to radicals generated in homogeneous solution. Thus, cyanoalkyl radicals generated by either photolytic or thermal decomposition of an azonitrile, or persulphate oxidation, or electrolysis of an α-cyanoacetic acid produces nearly the same ratio of C-C/C-N coupling products (Eq. (153)) [324].

$$NC-C(CH_3)_2CO_2^- \xrightarrow{S_2O_8^{2-}/-e}$$

$$NC-C(CH_3)_2-N=)_2 \xrightarrow{\Delta/h\nu} \quad NC-C(CH_3)_2^{\cdot} \longrightarrow \quad (153)$$

$$\underset{CH_3\ \ CH_3}{NC-\underset{\underset{CH_3\ \ CH_3}{|}}{\overset{\overset{CH_3\ \ CH_3}{|}}{C}}-\overset{}{C}-CN} + \underset{CH_3}{NC-\overset{\overset{CH_3}{|}}{C}-N=C=\overset{\overset{CH_3}{|}}{C}}$$

In sharp contrast, azonitrile adsorbed on silica produces only the C-C coupling product on photolytical decomposition [324a]. The Kolbe electrolysis of optically active carboxylic acids with the asymmetric carbon in the α-position yields inactive coupling products [335]. This result agrees with the postulate of a free radical as an intermediate, while for adsorbed radicals at least some retention of configuration would be expected.

Table 11 illustrates in some examples the scope of the Kolbe electrolysis.

Table 11. *Kolbe electrolysis of carboxylates*

Carboxylic acid	Product	Ref.
$CH_3(CH_2)_n COOH$ (n =0–14)	$CH_3(CH_2)_{2n}CH_3$ (40–90%)	326)
$ROOC-(CH_2)_n COOH$ (n = 1–16)	$ROOC(CH_2)_n - (CH_2)_n COOR$ (40–90%)	326)
$X(CH_2)_n COOH$ X=F, Cl, n = 4–10	$X(CH_2)_n -(CH_2)_n X$ (40–80%)	326, 336, 337)
$H(CF_2)_n COOH$ n = 4,6	$H(CF_2)-(CF_2)_n H$ (40–80%)	338)
$(EtO)_2CHCH_2CH_2-COOH$	$((EtO)_2CHCH_2CH_2)_2$ (60%)	339)
$Et\overset{O}{\overset{\|}{C}}-(CH_2)_4COOH$	$Et\overset{O}{\overset{\|}{C}}(CH_2)_8-\overset{O}{\overset{\|}{C}}Et$ (75%)	340)
$AcylNH(CH_2)_5COOH$	$AcylNH(CH_2)_{10}-NHAcyl$ (25–40%)	341)
1-Hydroxycyclohexyl-1-acetic acid	1,2-Bis(1-hydroxy-1-cyclohexyl)ethane (40%)	342)

Table 11 (continued)

Carboxylic acid	Product	Ref.
3-Acetoxybutyric acid	2,5-Hexanediol diacetate (59–81%)	343, 344)
6-Acetoxycaproic acid	1,10-Decanediol diacetate (83%)	344)
5-Acetoxyvaleric acid	1,8-Octanediol diacetate (80%)	344)

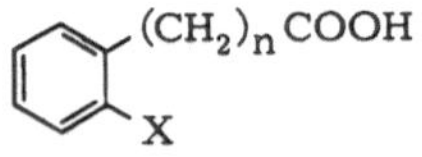

Dimer (3-30%)   345)

25% 346)

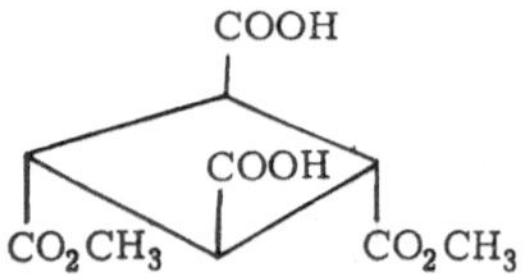

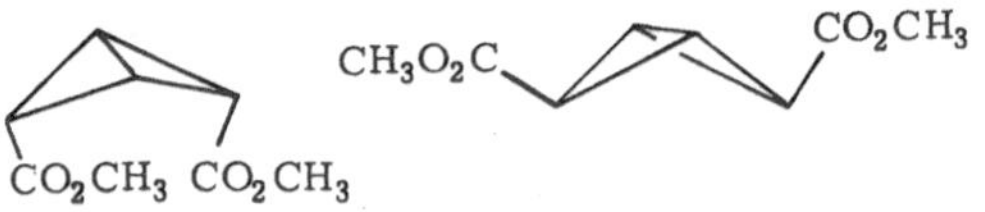

347)

Optically active centers in β to ω position of the carboxylic acid retain their full activity, just as double bonds in γ to ω position retain their *cis* or *trans* configuration.

A great variety of mixed couplings between different carboxylic acids have been performed to synthesize saturated branched, hydroxylated, unsaturated and optically active carboxylic acids, partially for the preparation of naturally occurring fatty acids [326]. Good yields of mixed coupling product were obtained if the less costly component was used in excess and both carboxylic acids had been fully neutralized to suppress preferential electrolysis of the stronger acid; see however Ref. [348a]. Some examples of mixed coupling are given below (Eq. (154–156)).

$$CH_3(CH_2)_6CO_2H + HO_2CCH_2\overset{\overset{\text{Me}}{|}}{C}HCH_2CO_2R \longrightarrow CH_3(CH_2)_7\overset{\overset{\text{Me}}{|}}{C}H\text{-}CH_2CO_2R$$

$$(154)\ [348]$$

$$\underset{\text{EtCH}}{\overset{\text{Me}}{|}} (CH_2)_m \, CO_2H + HO_2C(CH_2)_nCO_2R \longrightarrow \underset{\text{EtCH}(CH_2)_{m+n}-CO_2R}{\overset{\text{Me}}{|}} \qquad (155)\,^{349)}$$

$m = 1,2,4,6,8,14,18$

$n = 2,4,7,8,16$

$$CH_3(CH_2)_7C \equiv C \,(CH_2)_7CO_2H + \qquad\qquad (156)\,^{350)}$$
$$\qquad HO_2C(CH_2)_4CO_2Me \longrightarrow CH_3(CH_2)_7C \equiv C\text{-}(CH_2)_{11}CO_2Me$$

A great variety of substituted radicals for dimerization can be generated by anodic oxidation of anionic species $R^{\delta-}ME^{\delta+}$, *e.g.*, sodium salts of 1,3-dicarbonyl compounds, aliphatic nitro compounds, phenols, oximes, alkynes, thiolates or organometallics (Eq. (157)).

$$2\, R^{\delta-}-ME^{\delta+} \xrightarrow[\text{-ME}^+]{\text{-e}} 2\, R^{\cdot} \longrightarrow R\text{-}R \qquad (157)$$

Table 12 summarizes some representative examples.

Table 12. *Anodic dimerization of organic anions*

Anionic species	Product	Ref.
1,3-Dicarbonyl compounds:		
Diethyl sodio malonate (EtOH, DMSO, DMF)	Tetraethyl ethane-1,1,2,2-tetracarboxylate, hexaethyl propane-1,1,2,2,3,3-hexacarboxylate (main product in DMSO), tetraethyl 2-methylpropane-1,1,3,3-tetra-carboxylate	351)
Diethyl malonate (EtOH, CH$_3$CN, KI)	Tetraethyl ethane-1,1,2,2-tetracarboxylate (EtOH: 1,4 g/Ah, CH$_3$CN : 2,1 g/Ah)	352)
Triethyl sodiomethane-tricarboxylate	Hexaethyl ethanehexacarboxylate (40%)	354)
Ethyl acetoacetate (EtOH, CH$_3$CN/KI)	Diethyl diacetylsuccinate EtOH: 0,2 g/Ah, CH$_3$CN: 1,5 g/Ah	352)
Ethyl cyanoacetate	Diethyl α, α'-dicyanosuccinate	353)
Ethyl phenylacetate (EtOH, CH$_3$CN/KI)	Dimer: EtOH(trace), CH$_3$CN: 2,3 g/Ah	352)
Oximes:		
Potassium ethyl oximinopropionate (H$_2$O)	Diethyl 1,2-dinitroso-1,2-dimethylsuccinate	355)

101

Table 12 (continued)

Anionic species	Product	Ref.
Sodium diethyl oxi-minomalonate (H_2O)	sym. Dinitrosotetracarboethoxy ethane	355)

Aliphatic nitro compounds:

1-Nitropropane (H_2O, NaOH)	3,4-Dinitrohexane (25–30%), 3-nitro-3-hexene (45–55%)	356)
1-Nitrobutane (H_2O, NaOH)	4-Nitrooctene (35%), 4,5-dinitrooctane (35%)	356)
2-Nitropropane	2,3-Dinitro-2,3-dimethylbutane (15–45%)	357, 265)
2-Nitrobutane (H_2O, NaOH)	3,4-Dinitro-3,4-dimethylbutane (70%)	357)

Phenolates:

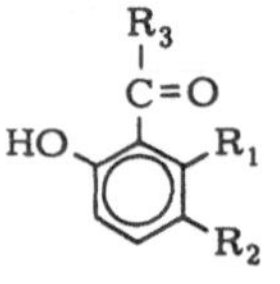

R_1:H, R_2:Me, R_3:Me 26% 358)

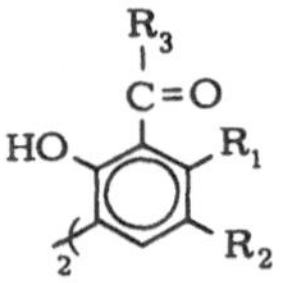

R_1:H, R_2:Me, R_3:Ph 17%

(H_2O, NaOH) Propylguajacol (H_2O, NaHCO$_3$)	Dipropyldiguajacol	358a)
2,6-Dihydroxyaceto-phenone (H_2O, NaOH, MeOH)	2,4,2',4'-Tetrahydroxy-3,3'-diacetylbiphenyl (52%)	358)
Corypalline		359)

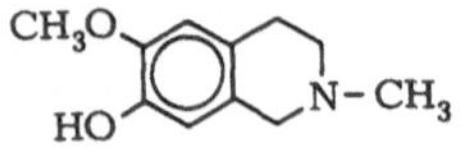

(28%)

(H_2O, NaHCO$_3$)

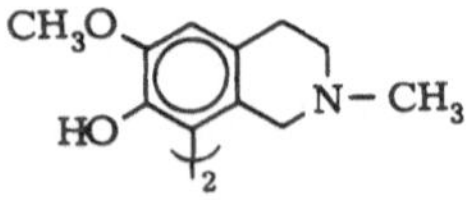

Alkynes:

$(CH_3)_2COH\text{-}C\equiv C^-$ $(CH_3)_2COH\text{-}C\equiv C\text{-})_2$ 360)

Table 12 (continued)

Anionic species	Product	Ref.
Thiolates:		
Sodium ethyl-mercaptide	Diethyl disulfide	361)
Sodium phenyl-mercaptide	Diphenyl disulfide	361)
Potassium thioacetate	Diacetyl disulfide	362)
Potassium isobutyl-xanthogenate	O-Isobutylthiocarbonic acid disulfide	363)
Organometallics:		
Isobutylmagnesium bromide (ether)	2,5-Dimethylhexane (94%)	364)
Amylmagnesium bromide (ether)	Decane (55−60%)	365)
Nonylmagnesium bromide (ether)	Octadecane (55−60%)	365)
Octadecylmagnesium bromide (ether)	Hexatriacontane (54%)	365)
Phenylmagnesium bromide	Biphenyl (55%)	366)
Diphenylzinc	Biphenyl (60%)	366)
Mixed coupling:		
Nitroethane (H_2O, KNO_2, Ag anode)	1,1-Dinitroethane (82%)	88)
2-Nitropropane (H_2O, NaOH, $NaNO_2$)	2,2-Dinitropropane	357)
Nitrocyclohexane (H_2O, NaN_3)	1-Azido-1-nitrocyclohexane (89%)	161)
2-Nitropropane (H_2O, NaN_3)	2-Azido-2-nitropropane	161)

Electrolysis of sodium salts of 1,3-dicarbonyl compounds often does not yield the wanted dimer, but ist methylene derivative *81* (Eq. (158)) [351,367] since the alcohol serving as solvent is oxidized to an aldehyde which reacts with the anion R⁻ by condensation and subsequent Michael addition (Eq. (158)). Cpe at a potential lower than the foot potential of the electrolyte should be applied in these cases.

$$\text{RCH}_2\text{OH (electrolyte)} \xrightarrow{-e} \text{R'CHO} \xrightarrow{\text{R}^-/\text{RH}} \text{R'CH=R} \xrightarrow{\text{R}^-} \underset{\text{R'}}{\text{R-CH-R}} \quad (158)$$

81

Alternatively, this unwanted side reaction is suppressed when electrolytes with higher positive foot potentials, *e.g.*, acetonitrile [352] DMSO, or DMF [351], are used. The dimer yields from nitroaliphatics are often diminished by β-eliminations of HNO_2 in the basic electrolyte, which leads to activated olefins [356] (Eq. (159)) that undergo follow-up reactions. The yields are generally better with secondary nitroaliphatics, where such eliminations are suppressed.

$$\text{CH}_3\text{CH}_2\text{CHNO}_2^- \longrightarrow \underset{\text{NO}_2}{\overset{\text{NO}_2}{\text{CH}_3\text{CH}_2\text{CH-CH-CH}_2\text{CH}_3}} \xrightarrow{-\text{HNO}_2} \underset{}{\overset{\text{NO}_2}{\text{CH}_3\text{CH}_2\text{CH=CHCH}_2\text{CH}_3}} \quad (159)$$

Oxidation of phenolates leads to dehydro dimers by coupling of two aryloxy-radicals in *o*- or *p*-position. Electrolysis of vanilline in acetonitrile-Et_4NOH yields 65% dehydrodivanilline (Eq. (160)) [280].

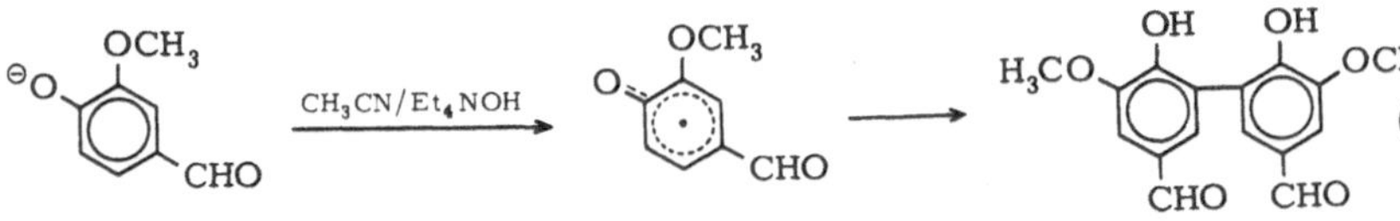

$$(160)$$

Analogously 2,4,2',4'-tetrahydroxy-3,3'-diacetylbiphenyl is obtained from 2,-6-dihydroxyacetophenone in water-methanol-NaOH in 52% yield (Eq. (161)) [358].

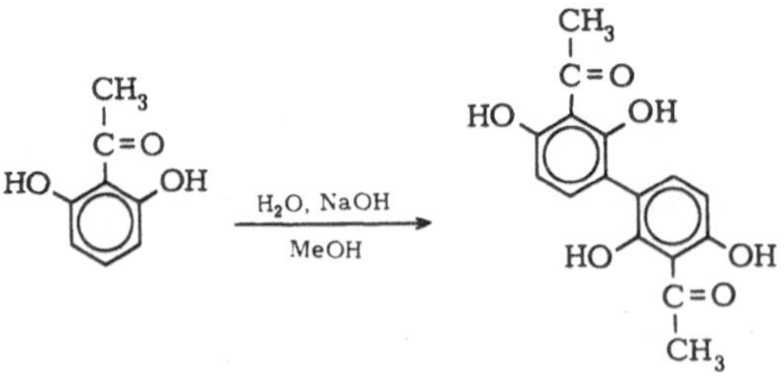

$$(161)$$

Yields of dimer in phenol oxidations are often decreased by competing reactions such as subsequent oxidation of the dehydro dimer or the anodically hydroxylated starting material to quinones which undergo 1,4-additions to higher molecular weight products. The wide choice of solvents and bases nowadays available should allow for suppressing these side reactions in suitable SSE's, such as $CH_3CN/NaClO_4/Et_4NOH$ [280].

Mercaptides [361] or xanthogenates [362,363] can be anodically coupled to the corresponding disulfides (Eq. (162)). A platinum anode impregnated with phthalocyanine has been used for mercaptide oxidation [368].

$$CH_3OC\overset{\overset{S}{\|}}{-}S^- \xrightarrow{-e} CH_3O\overset{\overset{S}{\|}}{C}-S-S-\overset{\overset{S}{\|}}{C}OCH_3 \qquad (162)$$

Grignard reagents [364–366a] or lithium alkyls serve as excellent sources for alkyl and phenyl radicals for anodic coupling. Due to the low discharge potential of organometallics, generally less than ± 0.0 V *vs.* the S.C.E., subsequent oxidation of α-branched alkyl radicals – as in the Kolbe electrolysis (Eq. (94)) – does not occur and good yields of dimer are usually obtained. Also the coupling of aryl groups, normally accomplished only in low yields *via* the Kolbe electrolysis, can be achieved satisfactorily by anodic oxidation of phenylmagnesium bromide, diphenylzinc or phenyl lithium. For Grignard electrolysis an anode process as in Eq. (163):

$$R_3Mg^-(R_2MgX^-, R_2Mg) \xrightarrow{-e} R^{\cdot} + R_2Mg(RMgX, RMg^+)$$
$$2\,R^{\cdot} \longrightarrow R\text{-}R \qquad (163)$$

and a cathode process as in Eq. (164):

$$MgX_2 \xrightarrow{+e} Mg + 2\,X^- \qquad (164)$$

is assumed [366a]. The electroactive species are continuously regenerated *via* the Schlenck equilibrium (Eq. (165)):

$$2\,RMgX \rightleftharpoons MgX_2 + R_2Mg \qquad (165)$$

To prevent short circuits by bridging of the electrodes due to magnesium deposits at the cathode, it is advisable to add an excess of the corresponding alkyl halide to the electrolyte to dissolve the magnesium deposits. The wealth of anions or organometallics [366b], providing a nearly inexhaustible supply of synthetic building blocks challenges anodic syntheses of dimers or mixed coupling products.

105

Efforts in this field of anodic oxidation are certainly to be expected. Difficulties that presently arise are due to the low conductivity in the usable solvents, *e.g.*, ether, tetrahydrofurane, diglyme, glyme, and the reactivity of the anionic precursors, which could lead to serious side reactions on prolonged electrolysis. These problems may possibly be overcome by low temperature electrolysis in capillary gap cells [366c] with small electrode distances to diminish the iR drop, and high electrode surface/electrolyte volume ratios for fast electrolysis.

Mixed coupling of two different anions, such as sodium azide or nitrite and a sodium salt of a nitroaliphatic compound, can be achieved by coelectrolysis of of both species. Product formation in these cases probably occurs by oxidation of N_3^- or NO_2^- to the corresponding radical that adds to the nitronate anion to form a radical anion that subsequently is oxidized to product (Eq. (168)) [161,357].

$$N_3^-(NO_2^-) \xrightarrow{-e} N_3^\cdot(NO_2^\cdot)$$

$$RC=NO_2^- + N_3^\cdot(NO_2^\cdot) \longrightarrow \overset{\overset{\displaystyle N_3(NO_2)}{|}}{RC} - NO_2^{-\cdot} \xrightarrow{-e} \overset{\overset{\displaystyle N_3(NO_2)}{|}}{RC} - NO_2 \qquad (168)$$

In an indirect electrolysis silver beads are used as anode in a flow cell to achieve coupling of sodium nitrite and nitroaliphatic in excellent yield (Eq. (169)) [88].

$$Ag^0 \longrightarrow Ag^+ \text{ (electrode process)}$$

$$2\,Ag^+ + NO_2^- + CR_2NO_2^- \longrightarrow 2\,Ag^0 + NO_2\text{-}CR_2\text{-}NO_2 \qquad (169)$$

b) Coupling of Anodically Generated Radical Cations

Aromatics, aromatic amines, phenols, and electron-rich olefins can be dimerized *via* radical cations according to Eq. (170).

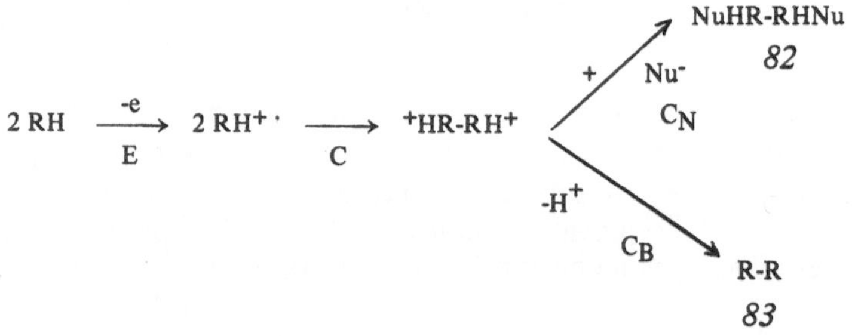

$$2\,RH \xrightarrow[E]{-e} 2\,RH^{+\cdot} \xrightarrow{C} {}^+HR\text{-}RH^+ \nearrow^{+Nu^-}_{C_N} \begin{matrix} NuHR\text{-}RHNu \\ 82 \end{matrix} \searrow_{-H^+}^{C_B} \begin{matrix} R\text{-}R \\ 83 \end{matrix} \qquad (170)$$

Thereby either substituted dimers *82* are formed in a ECC_N-sequence or dehydro dimers *83* in a ECC_B-sequence. Adams [25,29] has given a detailed discussion of these pathways, which needs not be duplicated here. Coupling of substrates *via* anodically generated radical cations is a powerful synthetic method. To illustrate this some representative examples from the wealth of available material are given in Table 13. More extensive compilations of data may be found in Ref.[10].

Table 13. *Anodic coupling via radical cations*

Substrate	Product	Ref.
Benzene (+ 2,4 V, CH_3CN)	Polyphenyl	369)
Benzene (HCl, $AlCl_3$)	Polyphenyl (90%)	370)
Mesitylene (HCl, $AlCl_3$)	Bimesityl (35%)	370)
o-Xylene (H_2O, acetone, H_2SO_4)	Bi-o-xylyl (39%) + other products	371)
1-Methylnaphthalene	4,4'-Bis (1-methylnaphthyl) (10%)	372)
1,2,4-Trimethoxy-benzene (H_2O, acetone, H_2SO_4)	2,4,5,2',4',5'-Hexamethoxybiphenyl (85%)	373)
Anthracene (EtOH, acetonitrile)	Bianthrone	374a, b)
Tetraphenylethylene (CH_3CN, $Et_4N^+ClO_4^-$)	9,10-Diphenylphenanthrene (100%)	375)

Aromatic amines:

Substrate	Product	Ref.
p-Nitroaniline (CH_3CN, pyridine, $NaClO_4$)	4,4'-Dinitroazobenzene	376)
2,4-Dichloroaniline	2,2',4,4'-Tetrachloroazobenzene (30%)	376)
Aniline	p-Aminodiphenylamine, benzidine	377)
9-Amino-10-phenyl-anthracene (aceto-nitrile, $LiClO_4$)	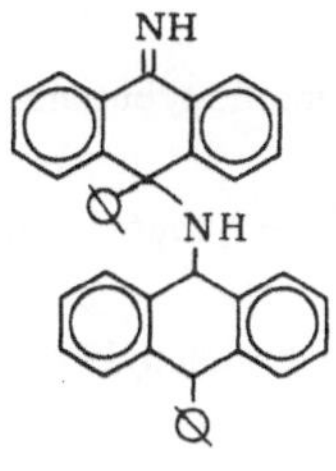	378)

107

Table 13 (continued)

Substrate	Product	Ref.
N-Methylaniline (cpe, H_2O)	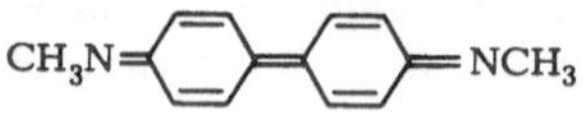 CH$_3$N=⟨⟩-⟨⟩=NCH$_3$	378)
N,N-Dimethylaniline	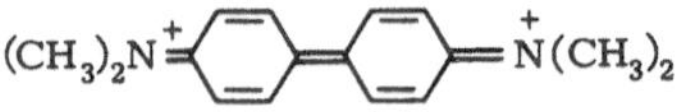 $(CH_3)_2\overset{+}{N}$=⟨⟩-⟨⟩=$\overset{+}{N}(CH_3)_2$ 2 X$^-$	379)
Triphenylamine (CH_3CN, $NaClO_4$)	N,N,N',N'-Tetraphenylbenzidine	380)
Carbazol	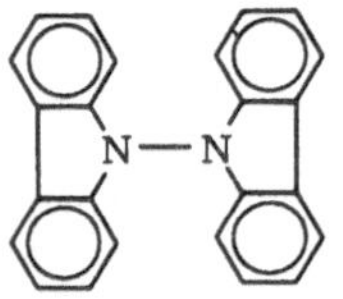	381)

Phenols:

p-Cresol (H_2O, H_2SO_4)	2,2'-Dihydroxy-5,5'-dimethylbiphenyl	382)
o-Cresol (H_2O, H_2SO_4)	3,3'-Dimethyl-4,4'-dihydroxybiphenyl	382)
3,5-Dihydroxybenzoic acid (H_2O, H_2SO_4)	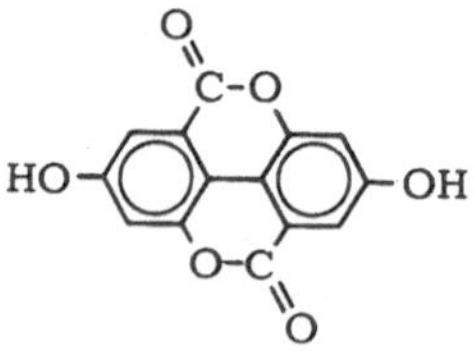	383)

Olefins:

Styrene (methanol, graphite)	1,4-Diphenyl-1,4-dimethoxybutane (60%)	36,268)
α-Methylstyrene (methanol, graphite)	1,4-Diphenyl-1,4-dimethylbutadiene (45%)	36,268)
Indene (methanol, graphite)	1,1'-Dimethoxy-2,2'-bisindenyl (38%)	36,268)

Table 13 (continued)

Substrate	Product	Ref.
4,4'-Dimethoxy-stilbene (cpe 1,05 V, CH₃CN, HOAc)	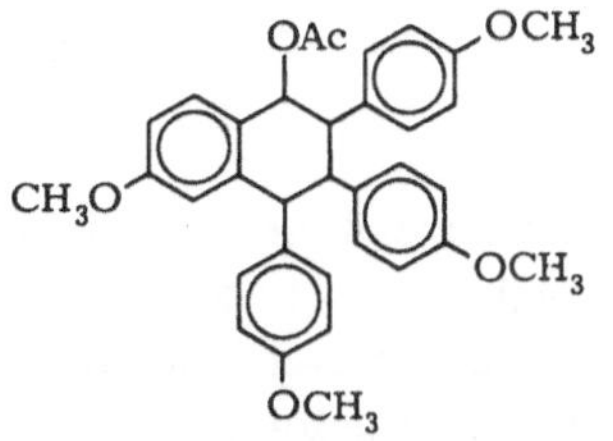	52)
Ethyl vinyl ether	Succindialdehyde dimethyldiethylacetal (51%)	35,36,268)
1-Ethoxy-1-cyclohexene	2,2'-Dicyclohexanone (27%)	36)
α-Ethoxystyrene	1,2-Dibenzoylethane (54%)	384)
Butadiene	trans, trans-1,8-Dimethoxy-2,6-octadiene (13%), trans-1,6-dimethoxy-2,7-octadiene (13%), 3,6-dimethoxy-1,7-octadiene (13%), dimethoxydodecatrienes (15%)	36,385)
1,1-Bis (dimethylamino)ethylene (CH₃CN, NaClO₄)	1,1,4,4-Tetrakis (dimethylamino) butadiene	386)

Mixed coupling of olefins:

Styrene-vinyl ethyl ether (methanol, NaI)	γ-Phenyl-γ-methoxyvaleraldehyde methylethylacetal (55%) [a]	384)
α-Methylstyrene-vinyl ethyl ether	γ-Phenyl-γ-methoxyvaleraldehyde methylethylacetal (32%) [b]	384)
α-Ethoxystyrene-vinyl ethyl ether	1,1,4-Trimethoxy-4-ethoxy-1-phenylbutane (45%) [a]	384)

a) Yield based on consumption of minor component styrene, b) current yield.

Benzene forms oligomers (biphenyl, terphenyl, sexiphenyl) whereas mesitylene yields bimesityl on electrolysis in a SSE consisting of a dimer complex $ArH \cdot HX \cdot Al_2X_6$. This interesting medium can dissolve over 50 mole-% aromatic hydrocarbon, has a specific conductance of $> 10^{-3}$ (ohm cm)$^{-1}$ and moderate to strong Friedel-Craft catalytic activity [370]. Anthracene is dimerized and subsequently oxidized to bianthrone [374a] in acetonitrile/ethanol by consumption of 3e/ mole anthracene, as was shown coulometrically (Eq. (171)) [374b].

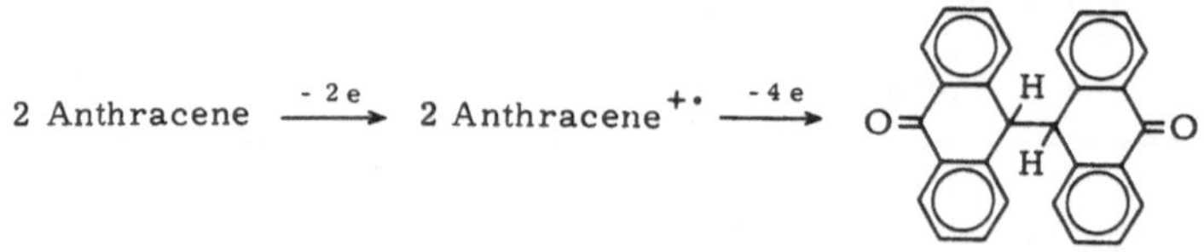

$$2 \text{ Anthracene} \xrightarrow{-2e} 2 \text{ Anthracene}^{+\cdot} \xrightarrow{-4e}$$ (171)

Tetraphenylethylene cyclizes anodically to 9,10-diphenylphenanthrene analogously to its photooxidative cyclization. The attempted anodic cyclization of *cis*- or *trans*-stilbene to phenanthrene however failed due to electrophilic reaction of the intermediate radical cation with the solvent [375]. Primary aromatic amines are oxidized to radical cations which, depending on the pH of the electrolyte couple to aminodiphenylamines (C-N coupling (*84*) in Eq. (172)), yield benzidines (*85*) at low pH (C-C coupling) or dimerize to hydrazobenzene (*86*) (N-N coupling) which is subsequently oxidized to azobenzene (Eq. (172)) [25,376,377].

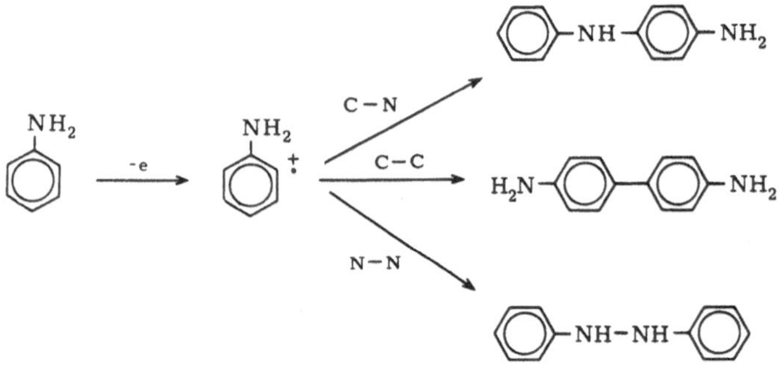

$$ (172) $$

Secondary and tertiary aromatic amines yield benzidines, which in some cases are subsequently oxidized to quinone imines *87* or benzidine dications *88* (Eq. (173)) [25,378-381].

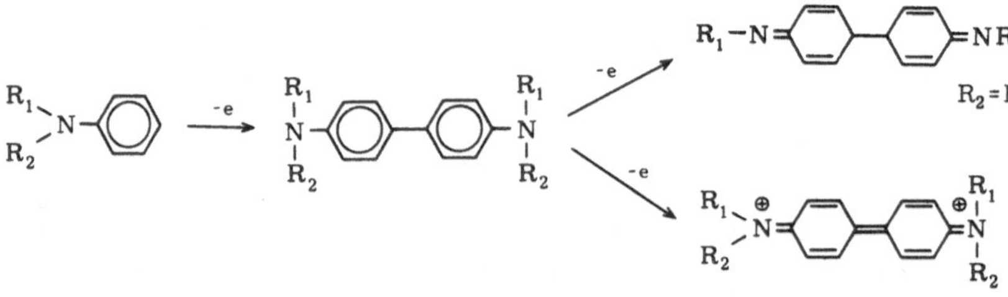

$$ (173) $$

The dimerization of butadiene, aryl olefins and ethyl vinyl ether is best rationalized by postulating a radical cation *89* (Eq. (174)) as first intermediate. As the β-carbon of *89* has the highest free valence, the highest positive charge density and the lowest atom localization energy radical or electrophilic reactions of *89* occur at the β-position. *89* either dimerizes (path a) to yield *90* and/or forms in an electrophilic addition (path b) to the olefin a 1,4-radical cation, which is subsequently oxidized to *90*. *90* produces substituted tail-to-tail dimers *91* or dienes *92* by S_N1-substitution or E1 -elimination.

$$
\begin{array}{c}
\underset{89}{\text{\Large$\diagup\!\!\!\diagdown$}} \overset{-e}{\longrightarrow} \underset{89}{\overset{Y \cdot ^{+}}{\text{\Large$\diagup\!\!\!\diagdown$}}} \\
\end{array}
$$

a) $\longrightarrow$ $\underset{90}{\overset{Y}{\oplus}C\text{-}C\text{-}C\overset{Y}{\oplus}} \overset{Nu^{\ominus}}{\underset{S_N1}{\longrightarrow}} \underset{91}{Nu\text{-}\overset{Y}{C}\text{-}C\text{-}C\text{-}C\text{-}Nu}$

b) $\longrightarrow$ $\underset{}{\overset{Y}{\cdot}C\text{-}C\text{-}C\text{-}C\overset{Y}{\oplus}}$

$\underset{-H^+}{\searrow}$ $\underset{92}{\overset{Y}{C}\text{=}C\text{-}C\text{=}C\overset{Y}{}}$

(174)

Anodic dimerization of electron-rich olefins, the mirror image process to cathodic hydrodimerization of activated olefins (Sect. 12.2), affords a one-step synthesis for substituted butanes (Eq. (175))[268], dienes (Eq. (176))[268], precursors of polyenes (Eq. (177))[36,385], and 1,4-dicarbonyl compounds (Eq. (178))[35,36].

$$CH_2=CH\text{-}C_6H_5 \xrightarrow[\text{NaClO}_4, \text{C}]{\text{-e/methanol}} \underset{}{C_6H_5\text{-}\overset{\text{OCH}_3}{CH}\text{-}CH_2\text{-}CH_2\text{-}\overset{\text{OCH}_3}{CH}\text{-}C_6H_5}$$

(175)

$$\underset{}{CH_2=\overset{\text{CH}_3}{C}\text{-}C_6H_5} \xrightarrow[\text{NaClO}_4/\text{NaOCH}_3/\text{C}]{\text{-e/methanol}} \underset{}{C_6H_5\text{-}\overset{\text{CH}_3}{C}\text{=}CH\text{-}CH\text{=}\overset{\text{CH}_3}{C}\text{-}C_6H_5}$$

(176)

$$CH_2=CH\text{-}CH=CH_2 \xrightarrow[\text{NaClO}_4, \text{C}]{\text{-e/methanol}}$$

$\underset{}{CH_2\text{-}CH=CH\text{-}CH_2\text{-}CH_2\text{-}CH=CH\text{-}CH_2}$ (OCH$_3$ ... OCH$_3$)

$\underset{}{CH_2=CH\text{-}CH\text{-}CH_2\text{-}CH_2\text{-}CH=CH\text{-}CH_2}$ (OCH$_3$ ... OCH$_3$)

(177)

$\underset{}{CH_2=CH\text{-}CH\text{-}CH_2\text{-}CH_2\text{-}CH\text{-}CH=CH_2}$ (OCH$_3$ OCH$_3$)

$\underset{}{CH_2\text{-}CH=CH\text{-}CH_2\text{-}CH_2\text{-}CH=CH\text{-}CH_2\text{-}CH_2\text{-}CH=CH\text{-}CH_2}$ (OCH$_3$ ... OCH$_3$)

$\underset{}{CH_2=CH\text{-}CH\text{-}CH_2\text{-}CH_2\text{-}CH=CH\text{-}CH_2\text{-}CH_2\text{-}CH=CH\text{-}CH_2}$ (OCH$_3$ ... OCH$_3$)

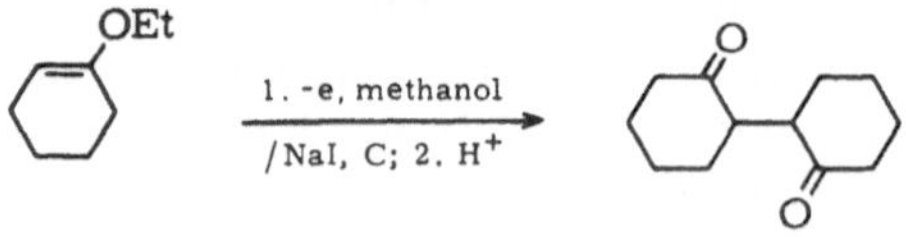

$$(178)$$

Mixed coupling of dissimilar olefins to acetals of substituted aldehydes (Eq. (179))[384] or unsymmetrical 1,4-dicarbonyl compounds (Eq. (180))[384] has also been achieved in fair to good yields.

$$(179)$$

$$CH_2=CHC_6H_5 + CH_2=CHOEt \quad \xrightarrow[C]{-e/methanol} \quad C_6H_5\text{-}\underset{\underset{OCH_3}{|}}{CH}\text{-}CH_2\text{-}CH_2\text{-}\underset{\underset{OCH_3}{|}}{CH}OEt$$

$$C_6H_5\underset{\underset{OEt}{|}}{C}=CH_2 + CH_2=CHOEt \quad \xrightarrow[C]{-e/methanol} \quad C_6H_5\text{-}\underset{\underset{OCH_3}{|}}{\overset{\overset{OCH_3}{|}}{C}}\text{-}CH_2\text{-}CH_2\text{-}\underset{\underset{OEt}{|}}{CH}\text{-}OEt$$

$$(180)$$

Diphenyldiazomethane is reversibly oxidized at a platinum electrode to a diazonium radical cation 93 (Eq. (181)) that reacts with excess diphenyldiazomethane 94 in a radical chain process to form tetraphenylethylene (95) as the main product and benzophenone (96) and benzpinacolin (97) as side products resulting from chain termination[387].

Initiation: $\quad (C_6H_5)_2C=N_2 \rightleftharpoons (C_6H_5)_2C=\overset{+}{N_2}\cdot$

$\qquad\qquad\qquad\quad$ 94 $\qquad\qquad\qquad$ 93

Radical chain: $\quad$ 93 + 94 $\xrightarrow{-N_2}$ $(C_6H_5)_2\overset{+}{C}$ —— $\overset{\cdot\cdot}{C}(C_6H_5)_2$ $\xrightarrow[-N_2]{94}$ 95 + 93

$\qquad\qquad\qquad\qquad\qquad\qquad\qquad\quad$ 99 $\underset{N_2}{\overset{||}{}}$

Termination: $\quad$ 93 + H_2O $\xrightarrow[-N_2]{-H^+}$ $(C_6H_5)_2\overset{\cdot}{C}OH$

$\qquad\qquad\qquad\qquad\qquad\qquad\qquad$ 98

$\qquad\qquad$ 99 + 98 $\xrightarrow{-H^+/-N_2}$ 96 + 95

$\qquad\qquad$ 93 + 98 $\xrightarrow{-H^+/-N_2}$ $(C_6H_5)_3C\text{-}\overset{\overset{O}{||}}{C}\text{-}C_6H_5$

$$(181)$$

$\qquad\qquad\qquad\qquad\qquad\qquad\qquad\quad$ 97

12.2. Cathodic Coupling

Cathodic coupling proceeds *via* radicals or radical anions, which are reductively generated from suitable substrates, in general electrophiles, *e.g.*, carbonyl compounds and activated olefins. These intermediates either dimerize (Eq. (182)) or add to activated double bonds to yield 1,4-radical anions, which are subsequently reduced to hydrodimers (Eq. (183)).

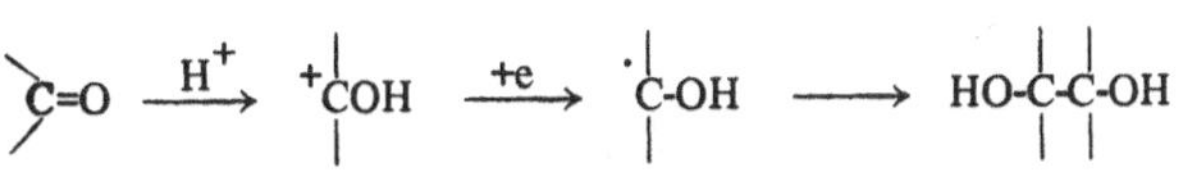

$$\text{C=O} \xrightarrow{\text{H}^+} {}^+\text{COH} \xrightarrow{\text{+e}} \text{C-OH} \longrightarrow \text{HO-C-C-OH} \qquad (182)$$

$$\text{C=C} \xrightarrow{\text{+e}} \text{C} \overset{-}{\underset{\text{Y}}{\text{C}}} \rightleftharpoons \text{C-C-C-C} \xrightarrow{\text{+e/H}^+} \text{H-C-C-C-C-H} \qquad (183)$$

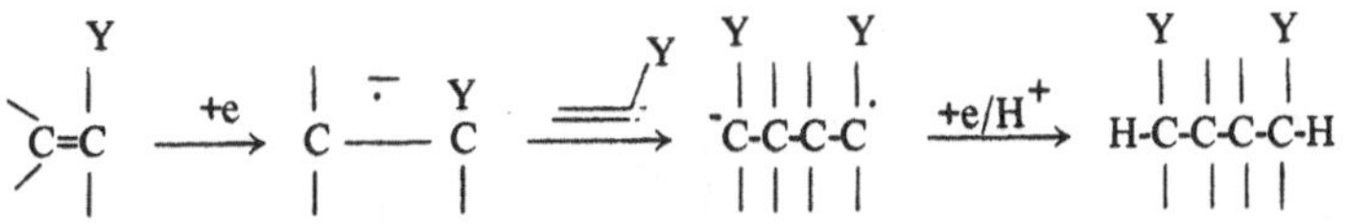

The subject has recently been reviewed by Baizer and Petrovitch [388] and the reader is referred to this excellent and comprehensive article for details. This section is subdivided into

a) the hydrodimerization of carbonyl compounds,

b) the hydrodimerization of activated olefins, and

c) the mixed hydrodimerization of dissimilar substrates.

a) Hydrodimerization of Carbonyl Compounds

The hydrodimerization of aromatic or aliphatic aldehydes and ketones yields pinacols (*100*) (Eq. (184)).

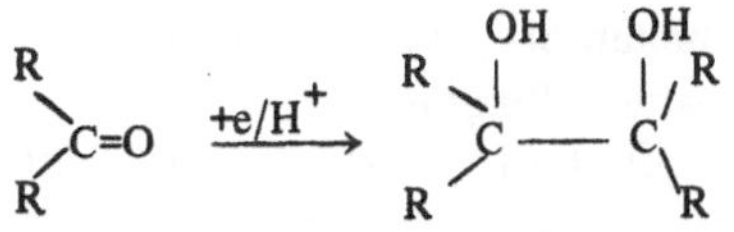

$$\underset{\text{R}}{\overset{\text{R}}{>}}\text{C=O} \xrightarrow{\text{+e/H}^+} \underset{\text{R}}{\overset{\text{R}}{>}}\underset{\text{C}}{\overset{\text{OH}}{|}} \text{—} \underset{\text{R}}{\overset{\text{OH}}{|}}\text{C}\overset{\text{R}}{<} \qquad (184)$$

100

113

The reduction mechanism of carbonyl compounds and its dependence on pH has been outlined in section 8.2. Pinacol formation occurs either by dimerization of the hydroxymethyl radical *101* (Eq. (185)) or by mixed coupling of *101* with the ketyl *102*. Dimerization of *102* seems less probable due to electrostatic repulsion of the two negative charges. Besides coupling, *101* or *102* may be further reduced to the alcohol *103*. With active cathodes (*e.g.*, Hg, Sn) *101* forms organo-

$$
\begin{array}{c}
\underset{101\ or\ 102}{\nearrow}\quad HO-\overset{|}{\underset{|}{C}}-\overset{|}{\underset{|}{C}}-OH \\
100
\end{array}
$$

$$
\cdot\overset{}{\underset{}{C}}-OH
$$

$$
101
$$

$$
+e/H^+/ME \qquad H-\overset{|}{\underset{|}{C}}-ME-\overset{|}{\underset{|}{C}}-H \xrightarrow[ME^{2+}]{H^+} H-\overset{|}{\underset{|}{C}}-H
$$

$$
104
$$

$$
H^+/+e
$$

$$
+e/H^+
$$

$$
\overset{}{\underset{}{C}}{=}O
$$

$$
H-\overset{|}{\underset{|}{C}}-OH
$$

$$
+e \qquad \overset{}{\underset{}{C}}-O^{-}\cdot \quad +e/2H^+ \qquad 103 \tag{185}
$$

$$
102
$$

metallics that are subsequently hydrolyzed to yield hydrocarbons *104*. The yield of pinacol depends on the successful suppression of these side reactions. With aliphatic carbonyl compounds pinacol formation is low to moderate in general, while alcohol and hydrocarbon formation predominates. Acetone has been dimerized at a zinc cathode to 34% pinacol [390] whereas with methyl ethyl ketone at Pb, Sn and Zn cathodes [394] or with cyclohexanone at Pb [391] moderate yields of the corresponding diols were obtained. The product ratio alcohol/hydrocarbon (*103*/*104*) depends critically on the cathode material. With acetone 67 to 95% isopropanol was isolated at Hg or Pb cathodes, besides mercury and organolead compounds, while with Cd, Zn, Al or Cu as cathode material a high yield of propane was obtained [389]. Cyclohexanone gave high yields of cyclohexanol at Zn and Cd cathodes and cyclohexane at Pb and Sn [391].

Pinacol yields are normally better with aromatic carbonyl compounds. Some examples are given in Table 14. A more extensive coverage - especially of the older literature - has been given by Allen [6], Popp and Schultz [9], and Fichter [4].

Table 14. *Hydrodimerization of carbonyl compounds*

Carbonyl compound	Electrode	Product	Ref.
Acetylbenzoyl	Cu, Sn	α, β-Diacetylhydrobenzoin (40%)	392)
Acetylbenzoyl	Hg	Organomercurial	392)
Fluorenone, tetralone		Pinacol, alcohol	393)
Benzophenone		Pinacol, benzhydrol	393, 401)
Acetophenone		α, α'-Dimethylhydrobenzoin, α-phenylethanol	393)
p-Aminoacetophenone	Hg, Sn	p,p'-Diamino-α, α'-dimethylhydrobenzoin (63%)	397, 398)
p-Hydroxypropiophenone	Hg, Pb	3,4-Bis (4-hydroxyphenyl)-3,4-hexanediol (90%)	395)
p-Hydroxybenzaldehyde	Hg	4,4'-Dihydroxyhydrobenzoin (95%)	399)
p-Dimethylamino-benzaldehyde	Hg	4,4'-Bis (dimethylamino)hydro-benzoin (58%)	396)
Furfural	Pb, Pt, Cu, Hg	1,2-Bis (3-furyl)-1,2-ethanediol (16−63%)	400)
2-, 3-, or 4-Acetyl-pyridine	Hg	2,3-Bis (2-, 3-, or 4-pyridyl)butane-2,3-diol (68−98%)	396, 409)
Benzaldehyde		Hydrobenzoin	402)
Benzoin	Hg	Pinacol (45%)	402, 403)
Desoxybenzoin	Hg	Pinacol (20−40%)	403)
Desylamine	Hg	Pinacol (30−40%)	403)
o-Vanillin	Pb	2,2'-Dihydroxy-3,3'-dimethoxy-hydrobenzoin (70%)	408)
p-Acetamidobenzal-dehyde	Hg	4,4'-Bis(acetamido)hydrobenzoin (80%)	401)

Available data on pinacol formation are not consistent enough to allow firm prescriptions on conditions for optimal yields. Pinacols are obtained at low and high pH, though a low pH may be more favorable. The dependence of yields on the cathode metal and the SSE are not yet fully understood. Too low and too high a pH are to be avoided to suppress nonelectrolytic pinacol rearrangements or aldol condensations as side reactions.

Jenevein and Stocker [132,404)] have studied the ratio of *d,l-* to *meso-*pinacols in the hydrodimerization of benzaldehyde and substituted acetophenones at different cathodes (Hg, Cu, Sn), current densities, cathode potentials and pH. The yields ranged in acid as well as basic medium between 40% to 90%. The *meso/d,l* ratio for acetophenone pinacol is independent of all parameters but basicity, which is rationalized as follows: In acidic medium two *101* (Eq. (185)) dimerize with no favor for the *d, l-* or *meso*-pinacol, in basic medium however *101* couples with *102*. Due to hydrogen bonding the two radicals become oriented in configuration *105* and production of the *d,l*-form is favored.

Cyclic glycols are obtained in moderate yields by intramolecular reductive coupling of the ketones *106* and *107* at a mercury cathode [410].

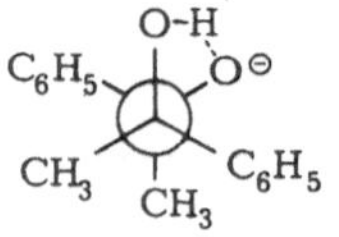

105 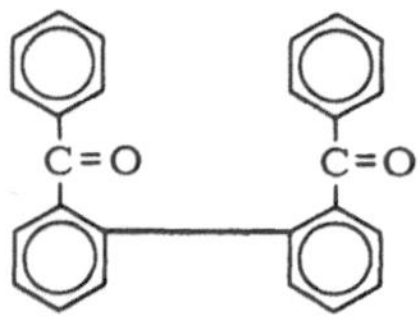   *106* 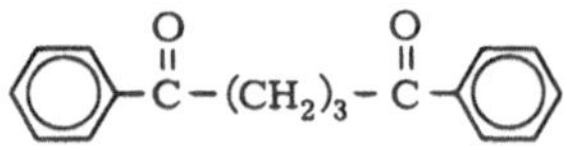   *107*

An interesting route to cyclopropanediol derivatives *via* intramolecular cyclization of properly substituted 1,3-diketones was opened by Curphey *et al.* [405, 406]. The 1,3-diketones *108* and *109* (Eqs. (186, 187)) were reduced at a mercury cathode in tetrahydrofuran or acetonitrile-$Bu_3EtN^+BF_4^-$ and in the presence of acetic anhydride to trap the intermediate diolates and prevent nonelectrolytic anionic ring cleavage. The cyclopropanediol diacetates *110* and *111* were obtained in 33% and 71% yield, respectively.

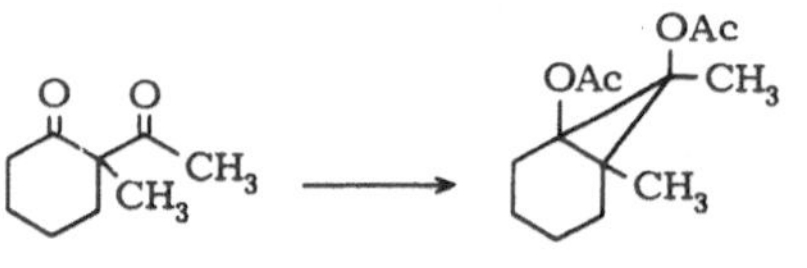

$$ \text{(186)} $$

108   *110*

$$ \text{(187)} $$

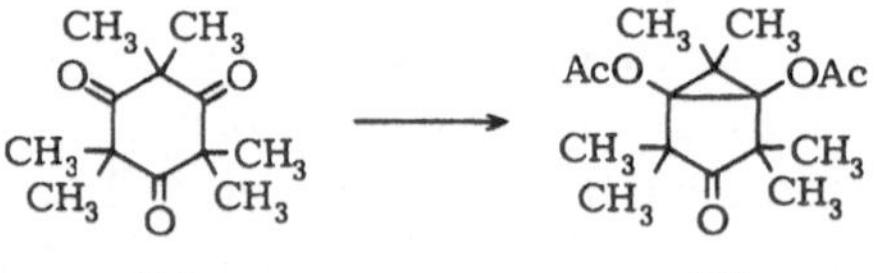

109   *111*

Schiff bases of substituted benzaldehydes can be hydrodimerized at the mercury cathode in ethanol-ethyl acetate - water - $Bu_4N^+Br^-$ in 25% to 66% yield to the diastereomeric 1,2-diamino compounds [407]. Coelectrolysis of a Schiff base with a carbonyl compound forms aminoalcohols.

116

b) Hydrodimerization of Activated Olefins

Activated olefins $\overset{\diagdown}{\underset{\diagup}{C}}=\overset{Y}{\underset{\diagdown}{C}}$ with Y: CN, C=O, COOR, $CONH_2$ can be cathodically coupled to tail-to-tail hydrodimers (Eq. (188)):

$$2\ \overset{\diagdown}{\underset{\diagup}{C}}=\overset{/}{\underset{\diagdown}{C}} \xrightarrow{+2e/2H^+} \text{H-C-C-C-C-H} \quad\overset{Y\ \ Y}{\underset{}{|\ |\ |\ |}} \tag{188}$$

Two short reviews on this powerful synthetic method are available [411, 423].

Cathodic coupling of α, β unsaturated carbonyl compounds yields three types of dimers: head-to-head (112, Eq. (189)), tail-to-tail (113) and head-to-tail hydrodimers (114).

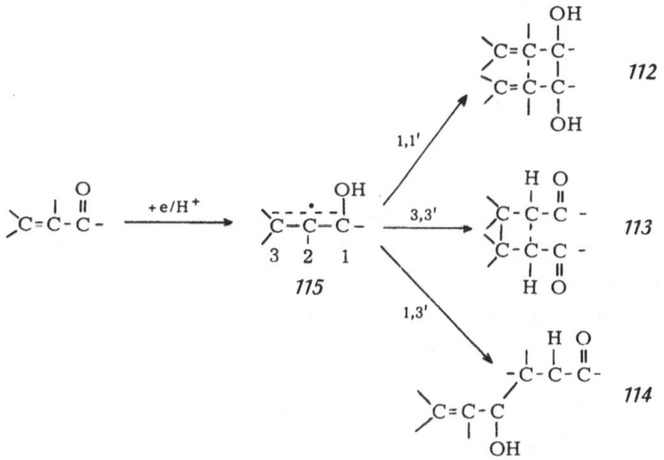

$$\tag{189}$$

These products are rationalized by assuming the hydroxyallyl radical 115 as the first intermediate, which is formed by 1e-reduction of the protonated α, β unsaturated compound. The hydrodimers 112 to 114 arise by 1, 1'-, 3,3'- and 1,3' coupling of 115. With methyl vinyl ketone a head-to-head and a tail-to-head dimer were obtained as side products [412] besides 70% 2,7-octanedione (product 113) as main product. At a pH <4 the organomercurial $Hg(CH_2CH_2COCH_3)_2$ is formed [413] Hydrodimerization of mesityl oxide yields four products, 116 to 119 [413, 414]. By proper choice of the reaction conditions each of them could be made the main product.

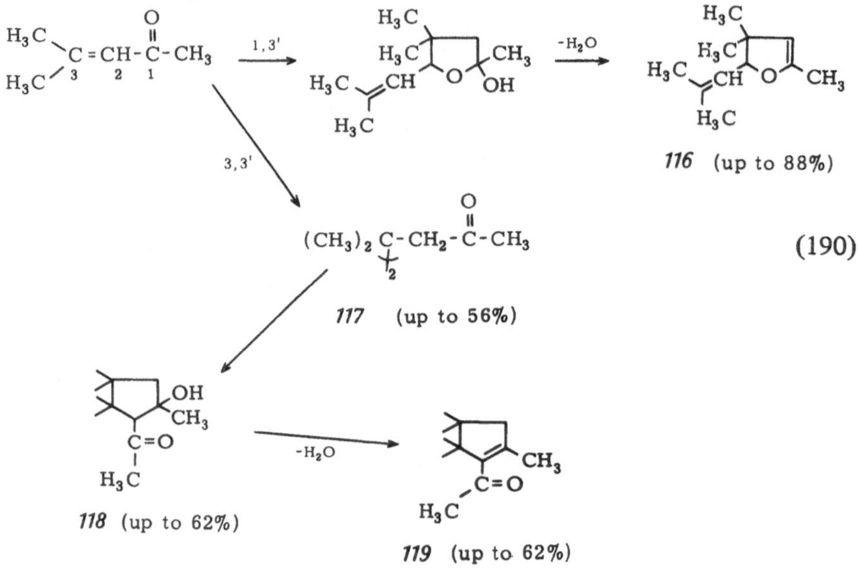

$$(190)$$

116 is formed by 1,3'-coupling of the intermediate allyl radical and subsequent intramolecular ketalisation and dehydratisation. 117 is the corresponding tail-to-tail hydrodimer. Intramolecular aldolisation of 117 forms 118, whose dehydratisation yields 119. With 2-cyclohexenone (Eq. (191)) the tail-to-tail (120) and the head-to-tail (121) hydrodimer were obtained [415,416]. The yield

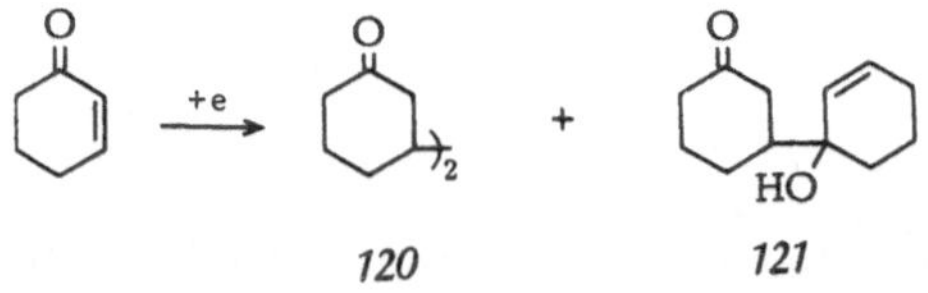

$$(191)$$

and product ratios were compared for both the chemical (Na/Hg, Mg, Zn) and electrochemical reduction. Yields and product ratios for electrochemical reduction are not very dependent on the solvent, while they vary considerably for chemical reduction (Table 15).

118

Table 15. *Electrochemical and chemical hydrodimerization of 2-cyclohexenone*

Solvent	cathode potential	electrochemical % yield	*120*	*121*	chemical % yield	*120*	*121*
HCl	-1,0 V	80	74	26	82	97	3
HOAc	-1,54 V	81	78	22	53	85	15
MeOH	-1,70 V	80	90	10	7	57	43

Bulk electrolysis of the crossconjugated dienone *122* (Eq. (192)) gave 96% of the pinacol *123* in acidic or basic medium. With the analogous methyltrichloromethyl derivative *124*, the pinacol yield dropped to 10%, but 76% of p-cresol was formed, probably by reductive elimination of the trichloromethyl group [417].

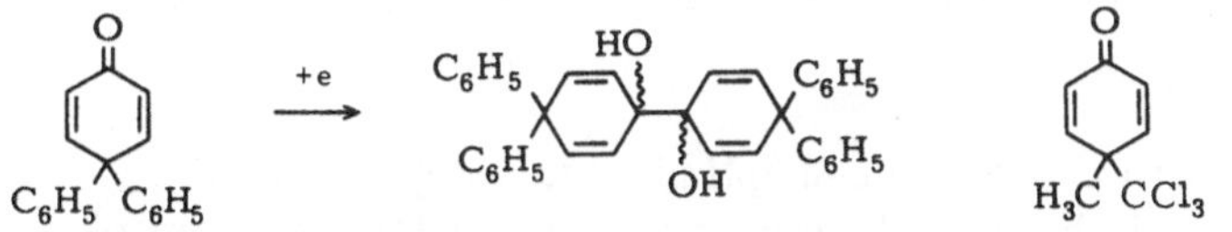

$$(192)$$

122   *123*   *124*

2,2,6,6-Tetramethyl-4-heptene-3-one (*125* in Eq. (193)) was reduced in dipolar aprotic media (DMF, HMTPA) to its radical anion *126* [418]. The ESR spectrum indicated 40–50% of the radical density at the β-carbon, and the rest at the carbonyl carbon and oxygen, but nothing at the α-carbon. *126* dimerizes to the tail-to-tail dimer at a rate which is the faster the more effectively the electrostatic repulsion of the two radical anions is reduced. With Li^+ as counterion the half life of *126* is less than 10^{-1} sec, with Na^+ or Pr_4N^+ 10 to 50 sec.

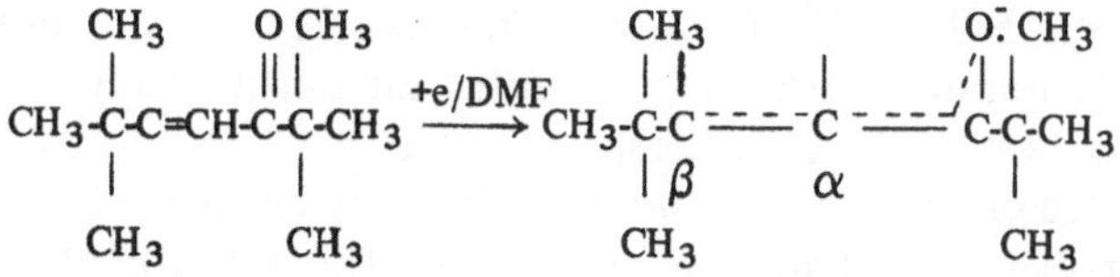

$$(193)$$

125   *126*

Baizer *et al.* [419] have compared the yields of hydrodimers from α, β-unsaturated esters obtained by chemical (=C: K/Hg, Li/Hg) and electrochemical (=E) reduction. In all cases the electrochemical method proved to be superior, *e.g.*,

ethyl acrylate: C 52% dimer, E 85%; diethyl maleate: C 31%, E 65% (Eq. (194));
acrylamide: C 0%, E 40%.

$$
\begin{array}{cc}
CO_2Et & CO_2Et \\
| & | \\
CH === CH
\end{array}
\xrightarrow{+e}
\begin{array}{cccc}
CO_2Et & CO_2Et & CO_2Et & CO_2Et \\
| & | & | & | \\
CH_2 \text{---} CH \text{---} CH \text{---} CH_2
\end{array}
\qquad (194)
$$

Hydrodimerization of ethyl acrylate, ethyl α-methylacrylate and ethyl croto-
nate yields exclusively tail-to-tail dimers [421], demonstrating regio selectivity of
cathodic coupling of α, β-unsaturated esters. Bulk electrolysis of the β, γ-unsa-
turated ester, ethyl 3-butenoate (Eq. (195)), forms a good yield of diethyl 3,4-
dimethyladipate, showing that compounds that are polarographically electroinac-
tive become reducible if "*in situ*" tautomerization leads to a conjugated system
[422]. Similarly allyl cyanide gives quantitatively 3,4-dimethyladiponitrile.

$$
CH_2 = CH\text{-}CH_2\text{-}CO_2Et \xrightarrow{+e/H^+}
\begin{array}{c}
CH_3\ CH_3 \\
| \ \ | \\
EtO_2C\text{-}CH_2\text{-}CH\text{-}CH\text{-}CH_2\text{-}CO_2Et
\end{array}
\qquad (195)
$$

Most extensively studied and optimized to nearly quantitative yield and current efficiency is the electrohydrodimerization (EHD) of acrylonitrile to adiponitrile [424]. This pioneering work of Baizer cannot be overestimated. Its outcome not only rendered the Monsanto process for the large scale production of
adiponitrile (40,000 tons/year) possible, but subsequent extension of the initial
work to EHD of a large series of activated olefins [425] made electrochemistry attractive as a synthetic tool and revived active interest in organic electrochemistry.
Besides the original Monsanto patent for EHD of acrylonitrile, numerous slightly
modified processes have been patented [426] later. Though EHD of acrylonitrile can
be conducted in comparable yield also with sodium amalgam [427], the electrochemical method (=E) proved much superior for the EHD of other α, β-unsaturated
nitriles: *e.g.*, α-methylacrylonitrile: chemical = C 0% dimer, E 75%; β,β-dimethyl-
acrylonitrile: C 0%, E 90% [419]. Contrary to the specific formation of tail-to-tail
dimers with α, β unsaturated esters crotononitrile and α-methylacrylonitrile give
small amounts of head-to-tail dimer as side product [435].

Other activated olefins that have been satisfactorily dimerized include:
1-cyano-1,3-butadiene to 1,8-dicyanoocta-2,6-diene (Eq. (196)) [428], 2-vinyl-
pyridine to 69% dimer, 4-vinylpyridine to 82% dimer (Eq. (197)) [429], α, β-un-
saturated phosphonates, phosphinates, phosphinoxides, and sulfones [430], and
aromatically substituted ethylenes [307,431].

$$
NC\text{-}CH\text{=}CH\text{-}CH\text{=}CH_2 \xrightarrow{+e/H^+} NC\text{-}CH_2\text{-}CH\text{=}CH\text{-}CH_2\text{-}CH_2\text{-}CH\text{=}CH\text{-}CH_2\text{-}CN \qquad (196)
$$

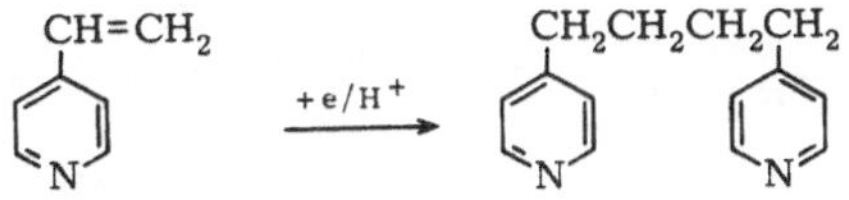

$$\text{(197)}$$

A preparatively interesting example of a combination of reductive coupling with subsequent reductive elimination is the formation of the alkyne *127* (Eq. (198)) from 1,1-bis (4-bromophenyl)-2,2,2-trichloroethane [431a].

$$(p\text{-Br-}C_6H_5)_2CH\text{-}CCl_3 \xrightarrow{+e} (p\text{-Br-}C_6H_5)_2\text{-}CH\text{-}C\equiv C\text{-}CH\text{-}(C_6H_5\text{-}pBr)_2 \quad \text{(198)}$$

$$127$$

Intramolecular reductive coupling of activated olefins = electrohydrocyclization, EHC, has been recently reviewed by Baizer and Petrovich [432]. The general reaction scheme of EHC is represented by Eq. (199):

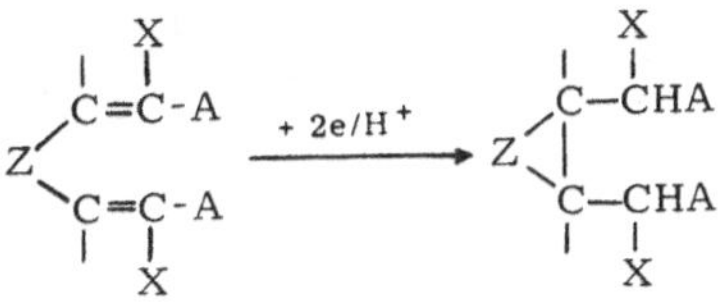

$$\text{(199)}$$

With A =H, X= CO_2 Et or CN and Z = CH_2 substituted cyclopropanes are formed in 98% yield; correspondingly with Z = $(CH_2)_2$ 15% cyclobutanes, with Z = $(CH_2)_3$ 100% cyclopentanes and with Z = $(CH_2)_4$ 90% cyclohexanes are obtained [433]. Attempted cyclizations to 8-, 10-, or 14-membered rings failed. *127a* cyclizes in 37% yield to the norbornene derivative *128* (Eq. (200)), *129* forms a 89% yield of the substituted 1,4-dioxane *130* (Eq. (201)), and *131* produces 86% of *132* (Eq. (202))[433].

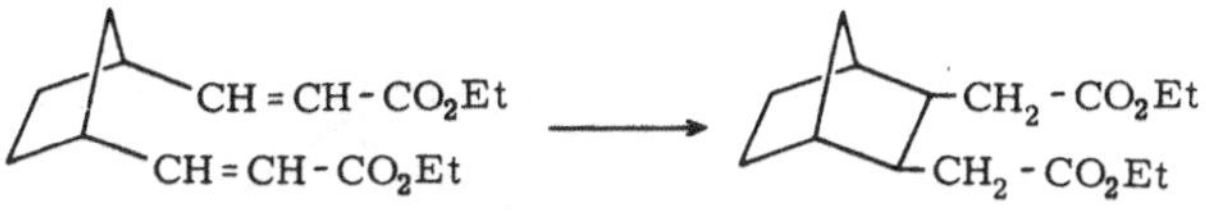

$$\text{(200)}$$

$$127a \qquad\qquad 128$$

$$\text{(201)}$$

$$129 \qquad\qquad 130$$

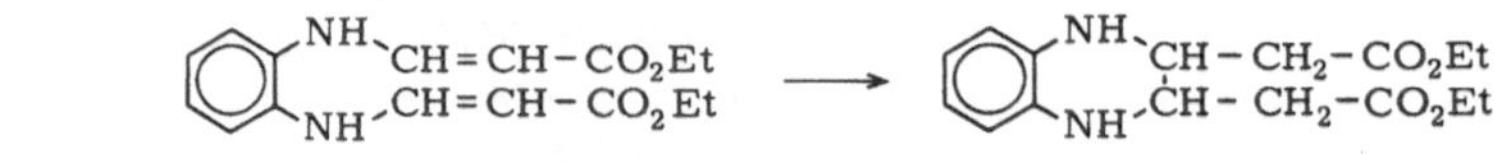

$$(202)$$

131   132

An intramolecular nucleophilic substitution of the dimethyl sulfide group by a cathodically generated radical anion is postulated in the formation of the indane derivative *133* from the sulfonium salt *134* (Eq. (203)).

$$(203)$$

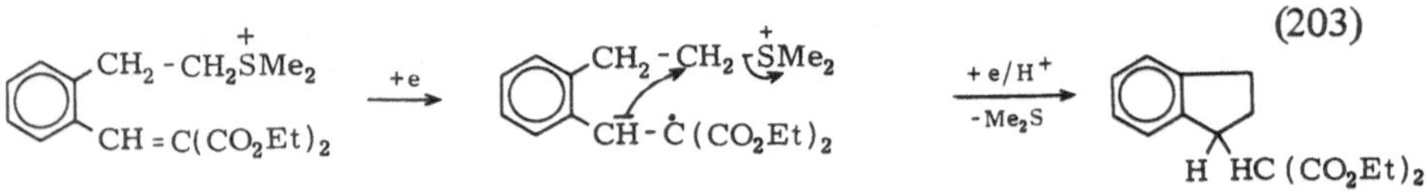

134   133

Cpe of 1,1'-ethylene-bis (3-carbamidopyridinium bromide) (*135* in Eq. (204)) in a 2e-reduction yields *136*, which can be reversibly cleaved at -0,25 V to *135*.

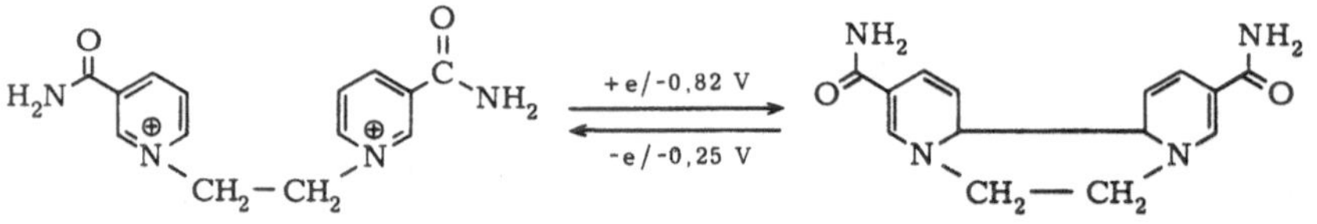

135   136 $$(204)$$

A detailed discussion of the mechanism of EHD is given by Baizer and Petrovich [388]. Originally Baizer [424,436] and Beck [437] proposed the following scheme for EHD, *e.g.*, of acrylonitrile (AN) (Eq. (205)):

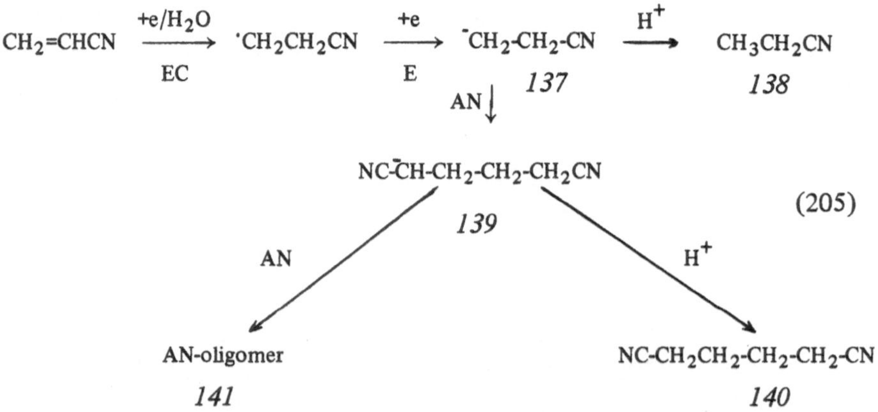

$$CH_2=CHCN \xrightarrow[EC]{+e/H_2O} \cdot CH_2CH_2CN \xrightarrow[E]{+e} \ ^-CH_2\text{-}CH_2\text{-}CN \xrightarrow{H^+} CH_3CH_2CN$$

                                              $AN\downarrow$ 137   138

$$NC\text{-}\overset{-}{C}H\text{-}CH_2\text{-}CH_2\text{-}CH_2CN$$

            139                             $$(205)$$

          AN H^+

AN-oligomer   NC-CH$_2$CH$_2$-CH$_2$-CH$_2$-CN

141   140

The anion *137* is formed by an ECE mechanism with the first electron transfer being rate determining. *137* may be protonated to propionitrile (PN) (*138*) or react with AN to *139* that is either protonated to adiponitrile (ADPN) (*140*) or initiates polymerization to AN-oligomers (*141*). Formation of PN is favored by high water concentration, ADPN by medium to low water concentration, while at very low water concentration oligomers are formed [438,457]. The ratio of ADPN/PN may be effectively controlled by the nature of the supporting electrolyte. Hydrophobic tetraalkylammonium cations, adsorbed at the cathode, deplete the electrode surface in water and favour ADPN formation, while with the hydrated Li^+ cation as supporting electrolyte predominantly PN is formed [80]. This phenomenon is also observed for EHD of α, β-unsaturated esters [420] and has been extensively discussed [80,439]. A two-electron reduction of AN with the dianion $\bar{C}H_2\text{-}\bar{C}H\text{-}CN$ as the key intermediate has been proposed by Asahera [440], Lazarov [441], and Tomilov [438]. Such a dianion with ET_4N^+ as counterion seems however relatively unlikely owing to electrostatic repulsion between the two negative charges. A third and most probable mechanism assumes reduction of AN to the radical anion *142* [442,443], which is either protonated to PN or reacts with AN to ADPN.

$$CH_2\text{=}CH\text{-}CN \longrightarrow \underset{\underset{142}{\big|}}{CH_2\text{-}\overset{-}{C}H\text{-}CN} \overset{H^+}{\longrightarrow} CH_3\text{-}\overset{\cdot}{C}H\text{-}CN \overset{+e}{\longrightarrow} \underset{143}{CH_3\overset{-}{C}HCN} \overset{H^+}{\longrightarrow} PN$$

$$\text{AN} \downarrow$$

$$NC\text{-}\overset{\cdot}{C}H\text{-}CH_2\text{-}CH_2\text{-}\overset{-}{C}H\text{-}CN$$

$$\downarrow +e/H^+ \tag{206}$$

$$\text{ADPN}$$

Protonation and addition of *142* occurs according to the atom localization energies calculated by Figeys [442], exclusively at the β-carbon. Neglecting field effects at the electrode the ECE sequence in Eq. (205) would yield the anion *143* and not *137*, whose addition to AN would produce head-to-tail dimers in contrast to the nearly exclusive formation of tail-to-tail dimers, which are readily explainable by the radical anion path (Eq. (206)). This mechanism is further supported by Baizer *et al.* [443]. They determined by cyclic voltammetry the reaction rates of radical anions generated from a series of bisactivated olefins $C(X)R = C(Y)R$ with X, Y = C_6H_5CO, C_6H_5, CN, CO_2Et, 4-Pyr. The rates of disappearance were proportional to the water and olefin concentration and the reaction with olefin was over 100 times faster than with water. The high reaction rate of the olefin was only partially attributed to the hydrophobic properties of the electrode surface induced by adsorbed tetraalkylammonium cations. For the other part the authors assume an induced reactivity of the olefin due to charge separation in the electric field (*143*). The radical anion reacts with the acceptor to a 1,4-radical anion with the anionic site outside the electric field. The radical site is further reduced to the anion and repelled from the electrode surface. With the bisactivated olefins three types of hydrodimers were obtained (*144, 145* and *146*).

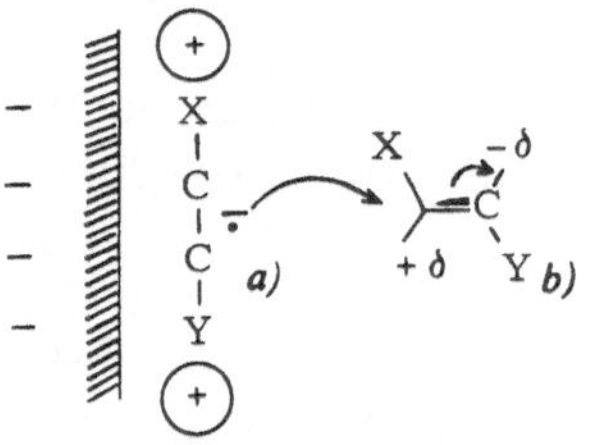

$$\begin{array}{ccccccc} & X & Y & Y & X \\ & | & | & | & | \\ H-&C-&C-&C-&C-&H \\ & | & | & | & | \end{array}$$

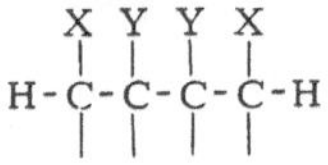

144

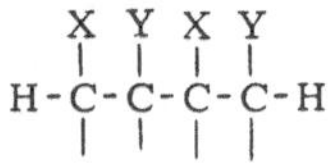

$$\begin{array}{ccccccc} & X & Y & X & Y \\ & | & | & | & | \\ H-&C-&C-&C-&C-&H \\ & | & | & | & | \end{array}$$

145

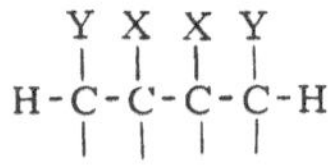

$$\begin{array}{ccccccc} & Y & X & X & Y \\ & | & | & | & | \\ H-&C-&C-&C-&C-&H \\ & | & | & | & | \end{array}$$

146

143

In the reaction of the radical anion (*143a*) with the acceptor olefin (*143b*) the 1,4-radical anion will be formed in which the activating groups in the acceptor best stabilize an α-anion and those in the donor best stabilize the α-radical. From this consideration the stabilizing efficiency of the substituents X,Y could be estimated from the product ratios *144/145/146* to be $C_6H_5CO > CO_2Et > 4$-pyridyl $> CN$ for the anionic site, and $C_6H_5CO > CO_2Et > C_6H_5 > CF_3$ for the radical site [444].

For EHC a concerted mechanism (Eq. (207)) is probable. The olefins which cyclize reduce at more positive anodic potential compared to those which do not cyclize. Ring formation is favoured by activating groups in the order $CN > CO_2Et > CONEt_2$ [445].

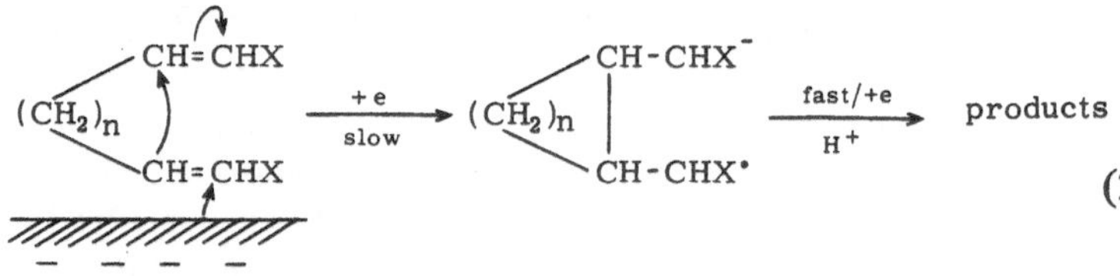

$$(CH_2)_n \begin{array}{c} CH=CHX \\ \\ CH=CHX \end{array} \xrightarrow[\text{slow}]{+e} (CH_2)_n \begin{array}{c} CH-CHX^- \\ | \\ CH-CHX^\cdot \end{array} \xrightarrow[H^+]{\text{fast}/+e} \text{products} \tag{207}$$

c) Mixed Cathodic Coupling

Three modes for the formation of cathodic mixed coupling products can be visualized:

1. *Mixed coupling of cathodically generated radicals* (Eq. (208)):

$$R_1R_2C=O + R_3R_4C=O \xrightarrow{+e/H^+} R_1R_2\overset{\cdot}{C}OH + R_3R_4\overset{\cdot}{C}OH \longrightarrow \overset{OH}{\underset{|}{R_1R_2C}}-\overset{OH}{\underset{|}{CR_3R_4}} \tag{208}$$

2. *Addition of cathodically generated radicals to double bonds* (Eq. (209)):

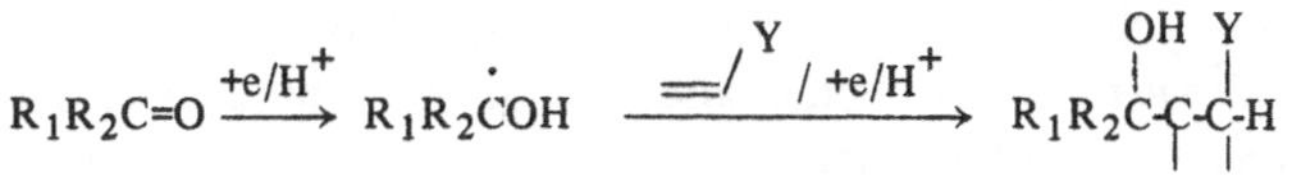

$$R_1R_2C{=}O \xrightarrow{+e/H^+} R_1R_2\overset{\cdot}{C}OH \xrightarrow{\underset{}{=}\diagup^Y / +e/H^+} R_1R_2\overset{OH}{\underset{|}{C}}{-}\overset{Y}{\underset{|}{C}}{-}\overset{}{C}{-}H \qquad (209)$$

Here the cathodically generated hydroxymethyl radical adds to the double bond to form a new radical, which is subsequently reduced and protonated to give product.

3. *Addition of cathodically generated radical anions to activated double bonds:*

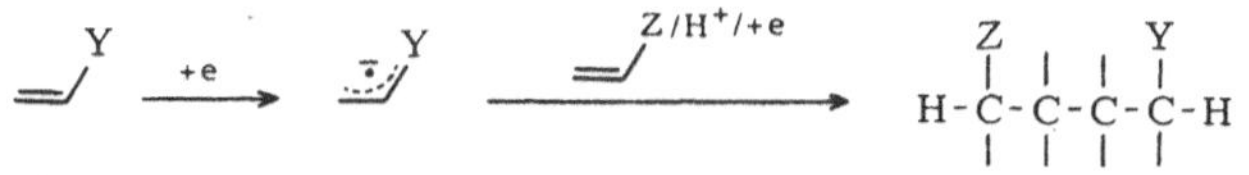

$$\underset{}{=}\diagup^Y \xrightarrow{+e} \underset{}{=}\diagup^Y \xrightarrow{Z/H^+/+e} \underset{}{=}\diagup^Y \longrightarrow H{-}\overset{Z}{\underset{|}{C}}{-}\overset{}{\underset{|}{C}}{-}\overset{}{\underset{|}{C}}{-}\overset{Y}{\underset{|}{C}}{-}H \qquad (210)$$

This route is essentially the radical anion mechanism (Eq. (206)) with donor and acceptor being dissimilar.

Path 1) is assumed for the mixed coupling of ketones to pinacols, for which examples are given in Table 16.

Table 16. *Mixed hydrodimerization of carbonyl compounds*

Ketone A	Ketone B	% Mixed pinacol	Ref.
p-Methoxyacetophenone	*p*-Dimethylaminoacetophenone	38	446,447)
p-Methoxyacetophenone	*p*-Aminoacetophenone	23	447)
Benzophenone	Acetone	18	451)

Sugino [448] obtained the crossed coupling product *147* in 70% yield and current efficiency on coelectrolysis of acrylonitrile and acetone in aqueous sulfuric acid at a mercury cathode. At lead and cadmium mixed coupling was suppressed and hydrocarbon formation increased. With methyl ethyl ketone and diethyl ketone crossed coupling was achieved in 60% and 30% yield, respectively. With acetone and maleic acid 10% terebic acid (*148*) was obtained. Tomilov [449] coupled acetone and acrylic acid in 95% yield (70% current efficiency) to

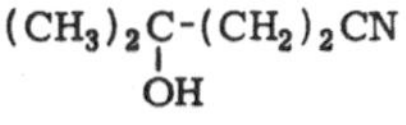

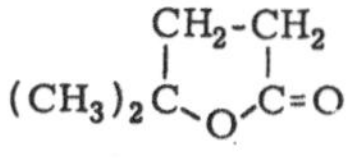

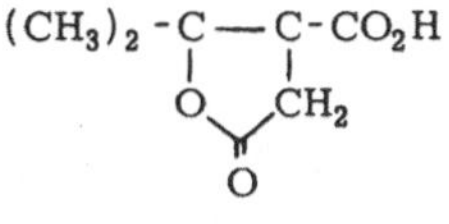

$$(CH_3)_2\underset{\underset{OH}{|}}{C}{-}(CH_2)_2CN$$

147

$$(CH_3)_2\underset{O}{\overset{CH_2-CH_2}{\underset{|}{C}}}\underset{}{\overset{|}{C}}{=}O$$

148

$$(CH_3)_2{-}C{-\!\!-}C{-}CO_2H$$

2-methyl-2,5-pentanediol. For the reductive coupling of acetone with acryloni-trile Brown and Lister [450] proposed path 2) as mechanism on the basis of pola-rization curves, cyclic voltammetry and capacitance data.

Probably also *via* path 2), Nicolas and Pallaud [451] obtained by coreduction of acetone and styrene 42% 2-methyl-4-phenylbutanol-2 (Eq. (211)), with ace-tone and butadiene 9% 2-methyl-5-hexene-2-ol and 23% 2-methyl-4-hexene-2-ol, and with acetone and stilbene 45% 2-methyl-3,4-diphenylbutane-2-ol (Eq. (212)). Similarly with acetone or benzophenone and ethyl acrylate dimethyl- (*149*) or

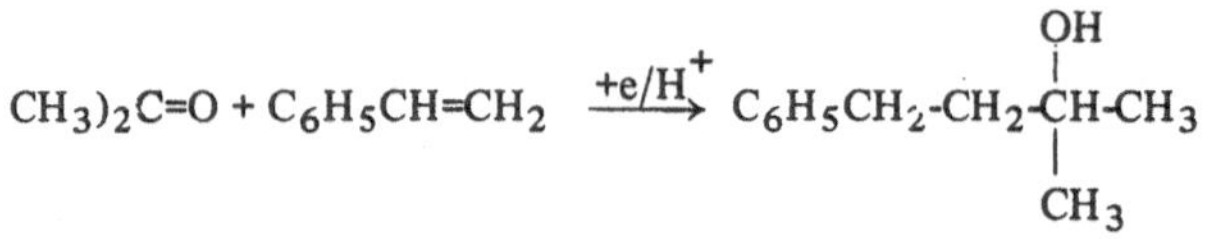

$$CH_3)_2C=O + C_6H_5CH=CH_2 \xrightarrow{+e/H^+} C_6H_5CH_2\text{-}CH_2\text{-}\underset{\underset{CH_3}{|}}{\overset{\overset{OH}{|}}{C}}H\text{-}CH_3 \qquad (211)$$

$$(CH_3)_2C=O + C_6H_5CH=CH\text{-}C_6H_5 \longrightarrow C_6H_5CH_2\text{-}\underset{\underset{CH_3}{|}}{\overset{\overset{C_6H_5}{|}}{C}}H\text{-}\overset{\overset{OH}{|}}{C}HCH_3 \qquad (212)$$

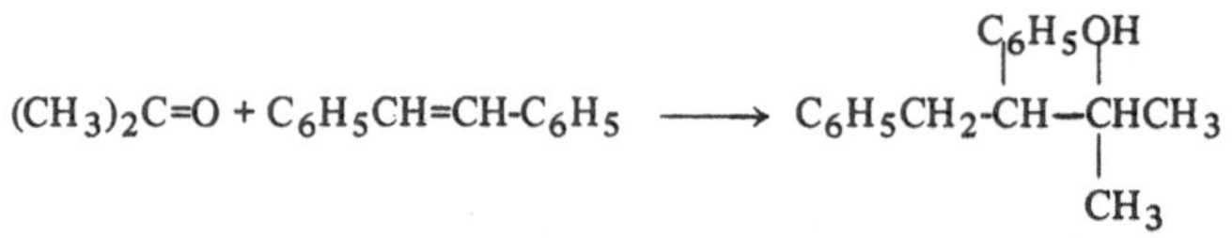

diphenylbutyrolactone (*150*) was obtained in 50% or 25% yield respectively. A mixed hydrodimer *151*, an additive dimer *152*, and an organomercurial formed in the coelectrolysis of triphenyl-β-cyanoethylphosphonium ion with styrene in-dicate a β-cyanoethyl radical as intermediate (Eq. (214)) [452].

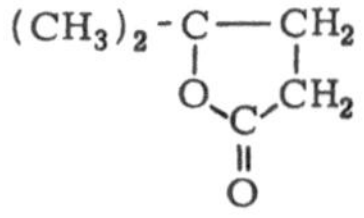

$$(CH_3)_2\text{-}\underset{\underset{\underset{O}{\|}}{\underset{O}{\overset{}{C}}}}{\overset{}{C}}\text{---}CH_2$$
$$\overset{}{O}\text{---}CH_2$$

149

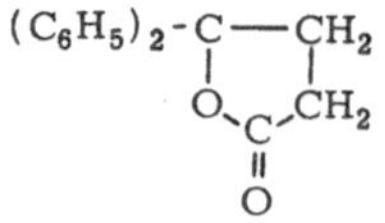

$$(C_6H_5)_2\text{-}\underset{\underset{\underset{O}{\|}}{\underset{O}{\overset{}{C}}}}{\overset{}{C}}\text{---}CH_2$$
$$\overset{}{O}\text{---}CH_2$$

150

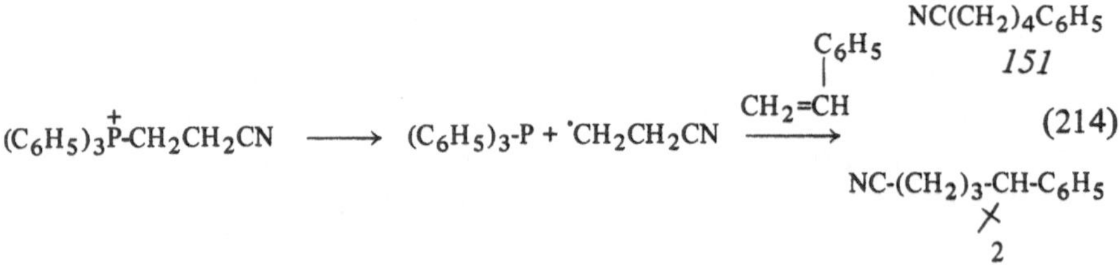

$$(C_6H_5)_3\overset{+}{P}\text{-}CH_2CH_2CN \longrightarrow (C_6H_5)_3\text{-}P + {}^{\cdot}CH_2CH_2CN \xrightarrow[\underset{CH_2=CH}{\overset{C_6H_5}{|}}]{} \begin{matrix} NC(CH_2)_4C_6H_5 & \quad & \textit{151} \\ \\ NC\text{-}(CH_2)_3\text{-}CH\text{-}C_6H_5 \\ \underset{2}{\times} \\ \textit{152} \end{matrix} \qquad (214)$$

Similarly mixed hydrodimers of type *151* were obtained with styrene and $(C_6H_5)_3\overset{+}{P}(CH_2)_nCN$ (n=3,4) [455]. Analogously the benzyl, cyanomethyl or carboethoxymethyl group in sulfonium or phosphonium salts can be transferred to acrylonitrile [455]. Pyridine and acetone have been coupled to mixed hydro-dimers in sulfuric acid solution [453].

Mixed coupling of two dissimilar activated olefins A and B is best rationalized by path 3). To suppress the formation of symmetric dimers AA and BB besides the wanted mixed dimer AB the difference in reduction potential between A and B should be 0,2 to 0,4 V. Cpe at the potential of the more easily reducible olefin A with an excess of B present in the electrolyte yields predominantly AB. With equal amounts of A and B AA and AB are obtained and with small diffe-rences in the reduction potentials of A and B all three possible dimers are for-med. Thus coreduction of diethyl maleate ($E_{1/2}$ = -1,32 V.) and acrylonitrile ($E_{1/2}$ = -1,94 V.) by cpe at -1,4 V yielded 153 (AA) and 154 (AB). Cpe at -1,7 V of 6 equivalents of AN and one equivalent of cyanobutadiene ($E_{1/2}$ = -1,5 V) produced 79% suberonitrile [454].

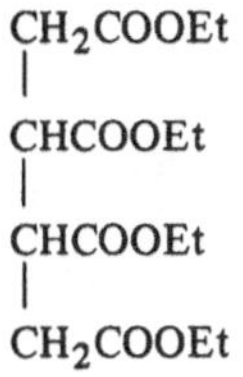

CH$_2$COOEt
|
CHCOOEt
|
CHCOOEt
|
CH$_2$COOEt

153

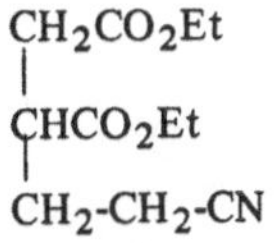

CH$_2$CO$_2$Et
|
CHCO$_2$Et
|
CH$_2$-CH$_2$-CN

154

Coelectrolysis of azo cpompounds and AN or ethyl acrylate yields 155 and 156, respectively. With 9-benzylidenefluorene and acrylonitrile a biscyanoethyl compound 157 was the major product [431]. Further mixed coupling products have been obtained with 2-vinylpyridine and dibutyl maleate [429] or methyl-vinylsulfone and N,N-diethylcinnamic acid amide [430].

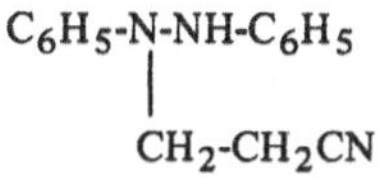

C$_6$H$_5$-N-NH-C$_6$H$_5$
|
CH$_2$-CH$_2$CN

155

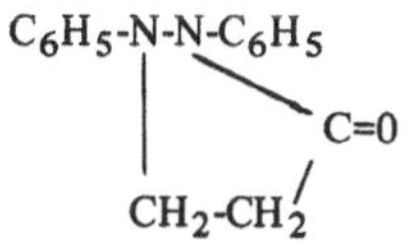

C$_6$H$_5$-N-N-C$_6$H$_5$
| \C=0
CH$_2$-CH$_2$/

156

CH$_2$-CH$_2$CN
CH-CH$_2$-CH$_2$-CN
|
C$_6$H$_5$

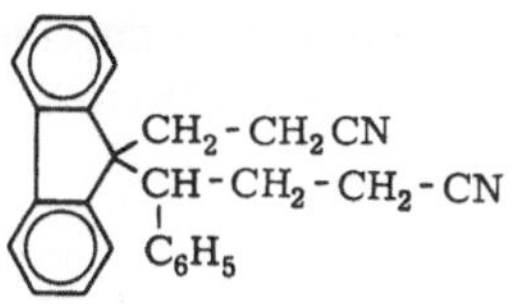

157

13. Electrochemical Cleavage

13.1. Anodic Cleavage

C-C or C-X bonds (X = I, NR_2, SR, OR) can be cleaved anodically by a) oxidation of the substrate to a radical cation that either dissociates (Eq. (215a)) or is hydrolytically decomposed in a subsequent C_BEC_N-process (Eq. (215 b)) and

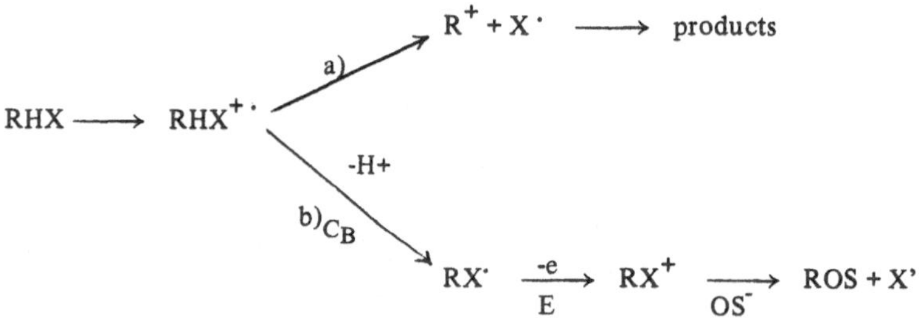

$$RHX \longrightarrow RHX^{+\cdot} \xrightarrow{a)} R^+ + X^\cdot \longrightarrow products$$

$$\xrightarrow[b)C_B]{-H+} RX^\cdot \xrightarrow[E]{-e} RX^+ \xrightarrow[OS^-]{} ROS + X' \tag{215}$$

b) by anodic generation of radicals, which subsequently fragmentate (Eq. (216)):

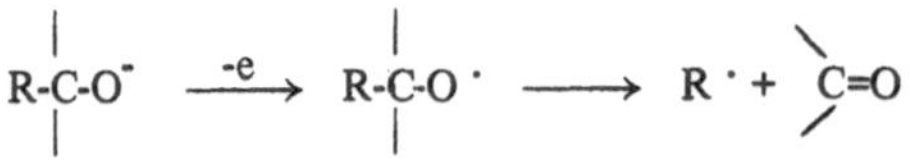

$$R\text{-}\underset{|}{\overset{|}{C}}\text{-}O^- \xrightarrow{-e} R\text{-}\underset{|}{\overset{|}{C}}\text{-}O^\cdot \longrightarrow R^\cdot + \diagdown C{=}O \tag{216}$$

Alkyl iodides are cleaved in acetonitrile/lithium perchlorate to iodine and alkyl cations. These react prior to or after rearrangement in a Ritter reaction with acetonitrile to form N-alkylacetamides in 21% to 75% yield (Eq. (217)) [458].

$$R\text{-}I \xrightarrow{-e} R\text{-}I^{+\cdot} \longrightarrow R^+ + 1/2 I_2 \xrightarrow{CH_3CN/H_2O} RNHCOCH_3 \tag{217}$$

In the analogous cleavage of 1,1-diphenyl-2-iodoethylene a vinyl cation is postulated as an intermediate whose follow-up reactions yield benzophenone and benzoin [459]. In contrast, aryl iodides are not decomposed anodically but form p-iododiaryliodonium salts via electrophilic substitution of the intermediate aryl iodide radical cation at iodobenzene (Eq. (218)) [458].

$$C_6H_5\text{-}J \xrightarrow{-e} C_6H_5\text{-}J^{+\cdot} \xrightarrow{C_6H_5\text{-}J} C_6H_5\text{-}\overset{\cdot}{\underset{H}{J}}\!\!\left\langle\!\!\!\!\!\begin{array}{c} + \end{array}\!\!\!\!\!\right\rangle\!\!\text{-}J \xrightarrow{-H^+,\,-e} C_6H_5\text{-}\overset{+}{J}\text{-}C_6H_4\text{-}J$$

$$(218)$$

1°-, 2°- and 3°-amides are cleaved in aquous acetonitrile to protonated dealkylated amides and aldehydes via immonium cations, which hydrolyze to product (Eq. (219), path a). In dry acetonitrile, however, the intermediate radical cation 158 abstracts hydrogen from acetonitrile to form the protonated amide and the α-cyanomethyl radical, the dimerization of which yields 83 - 95% succinonitrile (path b), Eq. (219)) [460].

$$(219)$$

Mizuno *et al.* [461] developed a synthesis for α, ω-dicarboxylic acids *via* path a), Eq. (219), by anodic cleavage of α, ω-N,N'-diacetyldiamines. As an example, with a PbO$_2$ anode in 1M sulfuric acid 21% malonic acid, 64% succinic acid, 51% adipic acid, or 46% pimelic acid was obtained on electrolysis of the corresponding N,N'-diacetylpolymethylenediamines. Similarly, N-acetylpolymethylenediamines were converted into amino acids [462], *e.g.*, N-acetylethylenediamine into 64% glycine, and N-acetylhexamethylenediamine into 44% ϵ-leucine.

The cleavage of primary amines in aprotic solvents, such as acetonitrile or DMF, was studied by cpe, controlled potential coulometry, and cyclic voltammetry [463]. The results indicate that at low anode potential the primarily formed radical cation $RCH_2NH_2^{+\cdot}$ dissociates to an alkyl cation RCH_2^+ and an amino radical $NH_2^{\cdot}$, which subsequently is oxidized to nitrogen. At higher anode potentials path a) of Eq. (219), leading to an aldehyde and an ammonium ion, preponderates.

Tertiary amines are dealkylated according to the high potential route to form secondary amines and aldehydes [464].

Aliphatic sulfides are oxidized to sulfones in SSE's containing water (see 8,1). In dry acetonitrile/sodium perchlorate, however, sodium methanesulfonate and CO is obtained. A methylthiomethyl cation 159, (Eq. (220)), is believed to be an intermediate, which is hydrolyzed to formaldehyde and methyl mercaptane. Both products are subsequently oxidized by Cl_2O_7, formed by dehydration of perchloric acid, to CO and $NaSO_3CH_3$ [465]. In spite of the fact that cpe was conducted well below the discharge potential of the supporting electrolyte complications arose (Eq. (220)), that were attributed to anodically generated $ClO_4\cdot$. This observation asks for caution in the use of perchlorates as supporting electrolytes in aprotic solvents. If possible, tetrafluoroborates or hexafluorophosphates should be used instead.

O-alkyl or O-acyl bonds are cleaved in the anodic conversion of 2,3,5,6-tetramethyl-1,4-dimethoxy- or -1,4-diacetoxybenzene to duroquinone [466].

Benzpinacol, fluorenepinacol, or xanthopinacol are cleaved at the mercury cathode to the corresponding ketones (Eq. (221)) [467].

$$CH_3\text{-}S\text{-}CH_3 \xrightarrow{-2e,\,-H^+} CH_2\overset{+}{=}S\text{-}CH_3 \xrightarrow{H_2O\text{ (from }HClO_4)} CH_2O \;+\; CH_3SH$$

$$159 \qquad\qquad \downarrow Cl_2O_7 \quad \downarrow \quad (220)$$

$$\underset{\substack{| \\ R}}{\overset{\substack{OH \\ |}}{R\text{-}C}}\!\!-\!\!\underset{\substack{| \\ R}}{\overset{\substack{OH \\ |}}{C}}\text{-}R \xrightarrow{-e/Hg} 2\,RRC{=}O \qquad (221) \qquad\qquad CO \qquad CH_3SO_3Na$$

Tertiary alcohols are oxidized in water-dioxane-NaOH to alkoxy radicals, which fragmentate to ketone and alkyl radicals $R\cdot$ (Eq. (216)). The relative rate of cleavage decreases with R in the order: sec -butyl $>$ isopropyl $>$ ethyl $>$ propyl $>$ pentyl $>$ isobutyl $>$ methyl [468]. Likewise, the bisulfite adduct of cyclohexanone is converted in 20% yield to 4-hydroxyhexanoic acid lactone (160) and 3-hydroxycyclohexanoic acid lactone (161) by anodic fragmentation (Eq. (222)) [469].

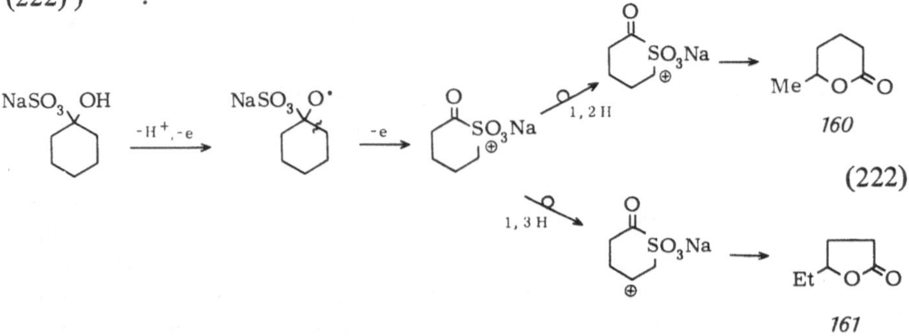

$$(222)$$

13.2. Cathodic Cleavage

The C-X or O-X'bond can be cleaved cathodically, whereby X,X' is substituted for hydrogen (Eq. (223)). Cathodic hydrogenolysis has been achieved with X =

$$-\overset{|}{\underset{|}{C}}-X \xrightarrow{+e/H^+} -\overset{|}{\underset{|}{C}}-H + X^-$$

$$-OX' \xrightarrow{+e/H^+} -OH + X'^-$$ (223)

$F, Cl, Br, I, SOR, SO_2R, NO_2, N^+R_3, P^+R_3, As^+R_3, SCN$ and CN, and $X' = COR, SO_2R$.

a) Cleavage of the C-Halogen Bond

The cleavage of the C-halogen bond has been most extensively studied. A fair amount of experiments have been designed to elucidate the reaction mechanism. For details the reader is referred to reviews by Elving [470], Elving and Pullmann [99], Perrin [15], and Brown and Harrison [12].

Three possible mechanism were proposed for the cleavage [470]: a) an S_N2-like process, b) an S_N1-like process, and c) a radical process.

In the S_N2- like process the cathode acts like a nucleophile in a S_N2 process. It attacks the substrate from the opposite side of the C-X bond, injecting two electrons and simultaneously displacing the halogen (Eq. (224)). The resulting carbanion is protonated by the SSE to product [477].

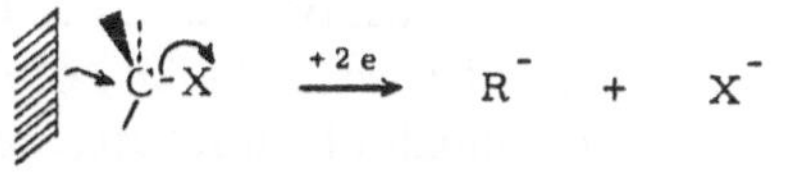

$$\overset{}{C}-X \xrightarrow{+2e} R^- + X^-$$ (224)

In the S_N1- like process the R-X bond ionizes in the strong potential field gradient $(10^7V/cm)$ [296] at the electrode/electrolyte interface and the carbonium ion is subsequently reduced to the carbanion (Eq. (225)). It has been suggested [15]

$$R-X \xrightarrow{+2e} R^+ + X^- \longrightarrow R^-$$ (225)

that mechanism a) and b) are essentially equivalent since when the carbonium ion is generated it will simultaneously draw electrons from the electrode.

In the radical process radicals R· are generated in a first 1 e-transfer and subsequently reduced in a second 1e-transfer to carbanions (Eq. (226)).

The radical path is most consistent with the existing experimental data. A correlation of half-wave potentials with Taft polar (σ^*) and steric (E_S) constants [478] indicates that a parallel orientation of the C-X dipole with regard to the electrode surface is most favourable for an electron transfer to the antibonding σ^* orbital of the C-X bond. Thereby a radical anion (162) is formed [471], which rapidly dissociates to halide ion and radical. The more facile reduction of exo-2-

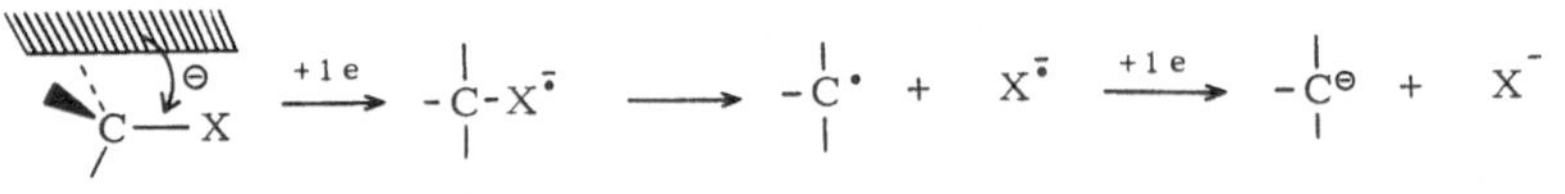

$$\overset{\text{\tiny ////////}}{\underset{/}{{>}\!C\!-\!X}}\;\xrightarrow{+1\,e}\; -\overset{|}{\underset{|}{C}}-X^{\bar{\cdot}}\;\longrightarrow\; -\overset{|}{\underset{|}{C}}\!^\cdot\; +\; X^{\bar{\cdot}}\;\xrightarrow{+1\,e}\; -\overset{|}{\underset{|}{C}}^{\ominus}\; +\; X^{-} \qquad (226)$$

bromide compared to the endo isomer or of cis-t-butylcyclohexyl bromide compared to the trans isomer supports this kind of steric approach [479]. The ease of reduction

$$I > Br > Cl > F, \text{ or } CX_3 > CX_2 > CX \;^{317)}$$

parallels the hypsochromic shift of the $p \longrightarrow \sigma^*$ transition in the UV-spectrum of the alkyl halides [472]. As expected for a transition state with radical character reduction is facilitated by R = allyl or benzyl and rendered more difficult by R = vinyl or phenyl [317]. Similarly the readiness of reduction decreases in the order $(CH_3)_2CBrCOOH > CH_3CHBrCOOH > CH_2BrCOOH$ [474]. 7-Norbornyl bromide is less easily reduced than adamantyl, triptycyl or C_2 to C_8 straight chain alkyl bromides [473], which reflects the decreased stability of the strained 7-norbornyl radical [476]. The half wave potentials of substituted benzyl chlorides are best correlated with a Hammett $(\sigma + \lambda \cdot \sigma \cdot) / \rho$ plot, which accounts for the radical anion character of the transition state [475]. A back-side displacement of X in an S_N2 like process seems improbable in view of the reducibility of bridgehead bromides or the relative small differences in the Taft steric parameters used to correlate the half wave potentials of various alkyl halides [478]. Also, attack on halogen is unlikely as no steric effects on the reduction potential should be noted in that case.

Though the validity of correlation of any homogeneous reaction kinetics with half wave potentials has been questioned [480], the nature of the cleavage products is most consistent with the radical mechanism. These products can be rationalized by the following scheme (Eq. (227)):

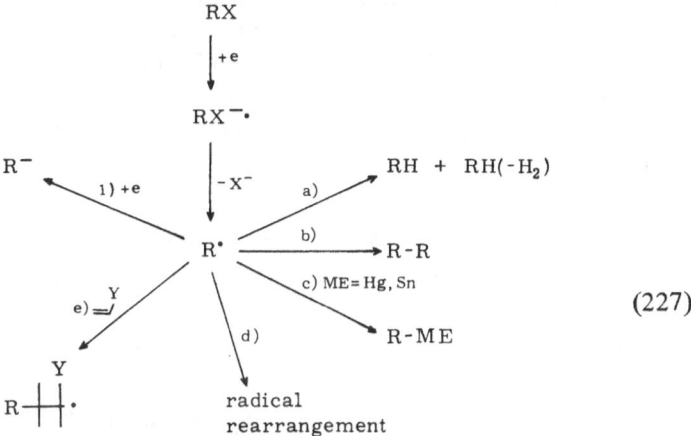

$$(227)$$

The initially formed radical R· disproportionates (path a),dimerizes (path b), reacts with active cathodes (path c), rearranges (path d), adds to double bonds (path e) or is reduced to an anion (path f). Products of radical origin (a–e) occur mainly in the reduction of alkyl iodides, benzyl halides and in some cases of alkyl bromides.

Reduction of β-iodopropionitrile at the potential of the first polarographic (1e) wave yields $Hg(CH_2CH_2CN)_2$, $Sn_2(CH_2CH_2CN)_6$, $Pb(CH_2CH_2CN)_4$ and $Tl(CH_2CH_2CN)_2I$ at Hg, Sn, Pb, or Tl cathodes [481]. When ethyl or butyl iodide is reduced at platinum or copper cathodes in DMF dimers R_2 and disproportionation products, RH, RH(-H$_2$), are formed [482,483,485]. Formation of these are attributed to radicals originating both from the cathodic cleavage of the R-I and the R_2-N(CH$_3$)$_2$ bond [482]. The ammonium salt is formed by alkylation of DMF.

Nitrobenzyl halides are reduced in a 1e-process to radical anions, which rapidly lose halide ion to form the neutral nitrobenzyl radical. The rates for this dissociation were calculated from cyclic voltammetry data to be $k = 2,5$ for m-nitrobenzyl chloride and $k = 2 \cdot 10^9$ sec^{-1} for o-nitrobenzyl bromide. The nitrobenzyl radicals predominantly dimerize (90%), whereas a small amount yields nitrotoluenes ($<10\%$) by hydrogen abstraction [484]. From a series of substituted benzyl bromides those with the more positive reduction potential form bibenzyl in 25–74% yield, whereas from the less easily reducible ones dibenzylmercury derivatives are obtained (50–60%) [485]. Reduction of benzyl chloride at the plateau of the first wave yields dibenzylmercury [486]. By reduction of diphenyliodonium hydroxide at -1,6 V 51% diphenylmercury is obtained [488]. In the reduction of (+)-S-1-bromo-1 -methyl-2,2-diphenylcyclopropane the intermediate cyclopropyl radicals are partially trapped by mercury to dicyclopropylmercury, which is subsequently reduced to the cyclopropanide ion. The configuration at Cl is only partially retained as the intermediate cyclopropyl radical

133

is configurationally labile. With the corresponding iodo derivative the dicyclo-propylmercury compound could be isolated [489].

Neophyl chloride in 13—75% current yield forms *t*-butylbenzene (94%) and isobutylbenzene (6%) as side product by rearrangement of the intermediate neophyl radical [487]. Reduction of monobromomaleic acid at pH 0—4 yields up to 50% of the dimer *163* (Eq. (228)). With increasing pH the amount of dimer

$$
\begin{array}{c}
\underset{Br}{\text{COOH}}\ \underset{H}{\text{COOH}} \quad \xrightarrow{+e} \quad \underset{2}{\left[\underset{}{\text{COOH}}\ \underset{H}{\text{COOH}}\right]} \\
163
\end{array}
\tag{228}
$$

decreases and maleic acid and fumaric acid preponderates [490]. In contrast tri-phenylbromoethylene is converted in a clean two-electron reduction to triphenyl-ethylene with no dimer formation [491]. Electrolysis of 6-bromo-1-hexene yields 80% hexene-1, besides dihexenylmercury and 7% methylcyclopentane by intra-molecular addition of the intermediate radical to the double bond [477]. These products of typical radical origin are formed when the reactions a) to e) in Eq. (227) are faster than the reduction of the radical (path f). With reduction poten-tials lower than -1.7 V (S.C.E.) these radical paths generally preponderate. With reduction potentials higher -2,0 V (S.C.E.), which applies to most alkyl chlorides and bromides, the second 1e-transfer is accelerated and generally anions are formed.

In protic electrolytes the solvent serves as proton donor for the anion. In aprotic solvents the anion either abstracts a β-hydrogen from the tetraalkyl-ammonium cation of the supporting electrolyte to form alkane, alkene and *t*-amine in a Hofmann elimination, or at low salt concentration the alkyl bromide is deprotonated in a β-elimination to yield a mixture of RH, and RH(-H$_2$) [477]. In the reduction of benzyl chloride the benzyl anion has been trapped in low yield ($<$1%) by CO$_2$ [305].

The stereochemistry of the C-X-bond cleavage ranges from partial retention for the cyclopropane derivative *164 via* partial inversion for *165* [489,492] to predominant inversion in the reduction of 2-phenyl-2-chloropropionic acid (*166*) to hydrotropic acid (*167*) [493].

164 : R :CH$_3$, COO$^-$

165 : R : COOH, CO$_2$Et

$$
\begin{array}{c}
\underset{C_6H_5}{\overset{R}{\diagup}}\underset{C_6H_5}{\overset{Br}{\diagdown}}
\end{array}
$$

$$
\underset{\underset{Cl}{|}}{\overset{\overset{C_6H_5}{|}}{CH_3-C-COOH}} \quad \longrightarrow \quad \underset{\underset{H}{|}}{\overset{\overset{C_6H_5}{|}}{CH_3-C-COOH}}
$$

166   *167*

Partial retention of configuration is attributed to anionic intermediates which slowly invert and thus are protonated with partially retained configuration. In the inversion case the anion is inverting more rapidly and is protonated from the side not shielded by the electrode thus yielding inverted product [492].

The difference in reduction potential for the C-Cl, C-Br and C-I bond can be profitably used for the selective reduction of polyfunctional molecules applying cpe [494]. Examples are:

$$1\text{-Br-4-(2-chloroethyl)-benzene} \xrightarrow{+e} (2\text{-chloroethyl)-benzene (99\%)}$$

$$m\text{-bromoacetophenone} \xrightarrow{+e} \text{acetophenone (94\%)}$$

$$p\text{-bromoiodobenzene} \xrightarrow{+e} \text{bromobenzene (98\%)}$$

$$p\text{-bromo-}\gamma\text{-chlorobutyro-phenone} \xrightarrow{+e} \gamma\text{-chlorobutyrophenone (96\%)}$$

A stereoselective cathodic hydrogenation has been achieved with *geminal* dihalocyclopropanes (*168*) (Eq. (229)). In all cases the *endo*-isomer *169* preponderated because of the more facile attack on *exo*-halogene and the stereoselectivity could be further enhanced by raising the water contents of the electrolyte.

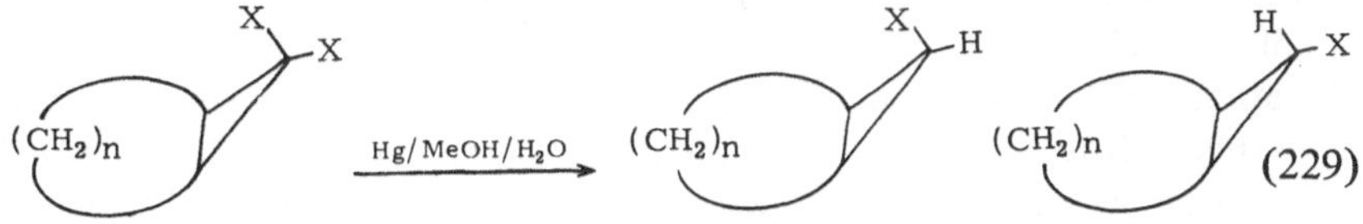

$$\text{Hg/MeOH/H}_2\text{O}$$

(229)

168 n: 3-4, X: Cl, Br   *169*

The trichloromethyl group can be reduced to either the dichloro- or the monochloromethyl group by proper choice of the SSE [495]. With ammonium nitrate the dichloro compounds are formed in 64—94% yield, while with tetramethylammonium chloride as supporting electrolyte the monochloro derivatives are obtained in 63 to 95% yield. As trichloromethyl compounds are readily available by telomerization of olefins with CCl_4, this reduction opens a simple route to α, ω-bifunctional aliphatic compounds.

Normally the C-F bond cannot be cleaved at the cathode. However, α to a carbonyl group or in a vinylogue position the C-F bond is readily hydrogenolyzed. α, α, α-Trifluoroacetophenone is transformed into 87% acetophenone [496], pentafluorobenzoic acid into 75% 2,3,5,6-tetrafluorobenzoic acid [497] and p-trifluoromethylbenzoic acid into 60% p-methylbenzoic acid [498]. Similarly a CF_3 group in o- or p-position to a sulfonyl group is readily hydrogenolyzed [499].

b) Cleavage of other Bonds

Electrolysis of quaternary ammonium salts in protic solvents, *e.g.*, methanol, cleaves the C-N bond with formation of a hydrocarbon and a tertiary amine [502].

The ease of reduction decreases in the order: benzyl $>$ allyl $>$ phenyl $>$ alkyl. Utilizing these differences in reducibility Horner *et al.* [505] opened an elegant route to unsymmetrically substituted amines $R_1R_2R_3N$ by combination of alkylation and cathodic cleavage.

With cyclic ammonium salts *170* (Eq. (230)) the ring opened amines are obtained in 90% yield on electrolysis in methanol at a mercury cathode [500]. Analogous ring openings are achieved with methiodides of N-methylindoline, N-

$$\tag{230}$$

170 $n = 2-4$

methyl-1,2,3,4,10,11-hexahydrocarbazole or N-methyl-1,2,3,4-tetrahydroisoquinoline on electrolysis in liquid ammonia [501].

Cathodic reduction of ammonium salts RR'_3N^+ in aprotic media, *e.g.*, DMF, yields dimers R-R for *e.g.*, R=benzyl, fluorenyl, in moderate yields (15-30%) [503]. Free radicals are assumed as intermediates, as inactive dimers were obtained from optically active ammonium salts *RR'_3N^+. This rules out a conceivable S_N2 reaction between cathodically generated R$^-$ and *RR'_3N^+ as products with partially inverted configuration should be obtained this way. Reduction of quaternary ammonium salts R_4N^+ in hexamethylphosphortriamide yields hydrocarbons RH and olefins RH(-H_2) as main products and dimers R-R as minor side products, whose formation is rationalized by assuming radicals R$\cdot$ as intermediates [504].

In phosphonium salts R_4P^+ the ease of cleaving the C-P bond to phosphine and hydrocarbon follows the order:

benzyl $>$ t-butyl $>$ isopropyl $>$ butyl $>$ ethyl $>$ phenyl $>$ methyl [506].

By applying alternative alkylation and cathodic cleavage Horner *et al.* [507] synthesized optically active phosphines and phosphonium salts. In reversal of their usual application as reagents in nucleophilic substitution halides, if converted to phosphoniumylides, can serve as nucleophiles in nucleophilic substitutions or additions. Combination of this reaction with reductive cleavage of the phosphonium salt recovers triphenylphosphine for a new cycle and forms the substituted product in good yield (Eq. (231)) [508].

(231)

$$RCH_2\text{-}X \xrightarrow[\text{2. B}^-]{\text{1.}(C_6H_5)_3P} (C_6H_5)_3\overset{+}{P}\text{-}\overset{-}{C}HR \xrightarrow{R'\text{-}X} (C_6H_5)_3\overset{+}{P}\text{-}CHRR' \xrightarrow{+e} RCH_2R'$$

Arsonium salts R_4As^+ can be cleaved with the ease of reduction in the following order:

benzyl $>$ allyl $>$ CH$_2$CO$_2$Et $> p$-CH$_3$-C$_6$H$_5$ $>$ C$_6$H$_5$ $>$ ethyl $>$ methyl [509].

As with the phosphines optically active arsines were obtained by combination of reduction and alkylation [510].

Amides and esters (Eq. (232)) afford amines and alcohols in a clean reaction on electrolysis at the mercury cathode in tetramethylammonium salt/methanol as SSE [511]. The tetraalkylammonium radical is believed to be the electron transfer reagent. With thioesters mercaptans are obtained in high yield [513].

$$C_6H_5\text{-}\overset{\overset{O}{\|}}{C}\text{-}NH\text{-}C_6H_5 \xrightarrow{+e} C_6H_5CH_2OH + C_6H_5NH_2$$

(232)

$$C_6H_5\text{-}\overset{\overset{O}{\|}}{C}\text{-}OCH_3 \xrightarrow{+e} C_6H_5CH_2OH + MeOH$$

Similarly, N-toluenesulfonamides are cleaved into amines and sulfinates. Peptides are cleanly desulfonated or debenzoylated without any cleavage of peptide bonds or racemization [511]. On electrolysis in liquid ammonia benzyl and carbobenzoxy protecting groups have been reductively removed from peptides [517]. Two dissimilar acyl groups, either in the same or different molecules, whose reduction potential differs by at least 0,2 V, can be selectively split off by cpe at the more positive reduction potential [512]. As an example, in a mixture of N-tosylanilide ($E_{1/2} \sim$ -2,23 V) and benzanilide ($E_{1/2} \sim$ -2,01 V) 96% of the benzoyl derivative is cleaved and only 4% of the tosyl derivative. On electrolysis of N-tosyl-N'-benzoyl-p-phenylenediamine 90% of N-tosyl-p-phenylenediamine is obtained. In this way Horner et al. [512] demonstrated that a selective removal of protecting groups in peptide synthesis is possible.

Electrolysis of a tosylate in methanol as electrolyte recovers the alcohol with retention of configuration by cleavage of the O-SO$_2$ bond [514]. In aprotic solvents (CH$_3$CN) this cleavage is not so clean, alcohol, ether and toluene being formed simultaneously [515]. Electrolysis of diaryl and arylalkylsulfones forms sulfinates and hydrocarbons in 75–90% yield by hydrogenolytic cleavage of the C-SO_2 bond. The reduction potential increases in the series benzyl $<$ alkyl $<$ aryl [516]. It is remarkable that also sterically very hindered sulfones, *e.g.*, bis-(2,4,6-trimethylphenyl) sulfone, are readily cleaved, which cannot be matched by any conventional reduction.

Functional groups α to a carbonyl group can be readily removed at the mercury cathode. Reduction of various α-substituted acetophenones X-CH_2-$\overset{O}{\overset{\|}{C}}$-$C_6H_5$ with $X = OAc$, OH, OC_6H_5, and SCN yields acetophenone [518]. Cleavage of bicyclic α-aminoketones affords an elegant synthesis for 9- to 12-membered hydroxyazacycloalkanes (Eq. (233)) [519,520].

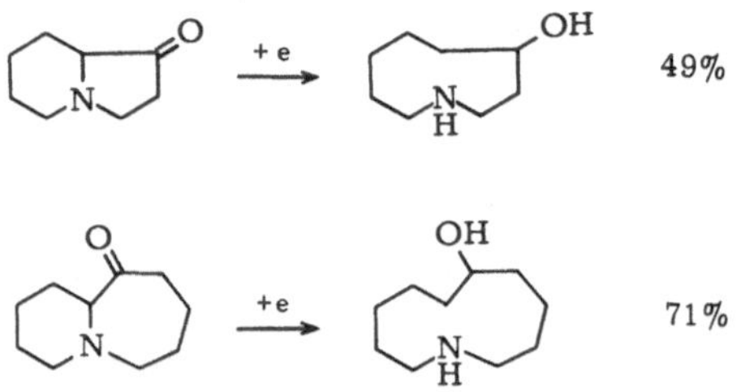

$$(233)$$

Ketosulfoxides, which are intermediates in the alkylation of ester by Li-CH_2SOCH_3, are cleaved to ketones (Eq. (234)) [521]. Analogously, the acetoxy group has been reductively removed from α-acetoxylated nitriles [522].

$$(234)$$

$$R = p\text{-}MeOC_6H_4, \text{ hexyl}$$

The cyano group in nitriles can be readily split off by electrolysis in anhydrous ethylamine. 1,1'-Dicyanodicyclohexyl, cycloheptanecarbonitrile, octanecarbonitrile or dehydroabietonitrile form the corresponding hydrocarbons in 60–80% yield [523]. The solvated electron is believed to be the reducing agent.

Benzyl ethers with electron attracting groups in the phenyl ring are cleanly converted (60–95%) to the corresponding toluenes [212,498].

From tertiary aliphatic nitro compounds the NO_2-group can be reductively removed. By ESR and cyclic voltammetry a radical anion has been demonstrated to be the first intermediate (Eq. (235)) [524]. This dissociates to $N\bar{O}_2$

$$R\text{-}NO_2 \xrightarrow{+e} RNO_2^{-\cdot} \longrightarrow R^{\cdot} + NO_2^{-} \qquad (235)$$

and a tertiary radical. The radical undergoes a number of follow up reactions (see also 14.4), such as coupling with the radical anion to nitroxides, disproportionation, dimerization to R_2, or cathodic reduction to R^-.

14. Electron Transfer

14.1. Generation of Radical Cations

One electron transfer from the highest filled MO of a neutral substrate *170* (Eq. (236)) to the anode yields a radical cation *171* as product. This may be either a transient intermediate or a stable, long-lived species depending on its substituents and the nucleophilicity of the solvent. The reaction paths of radical cations have been expertly and comprehensively reviewed by Adams [25,29], so that a short summary seems sufficient at this place. Deprotonation and 1e-oxidation of *171* with a subsequent S_N1 reaction of the resulting cation yields side-chain substitution products *172* (path a), see 9.1. Solvolysis of *171* followed by 1e-oxidation

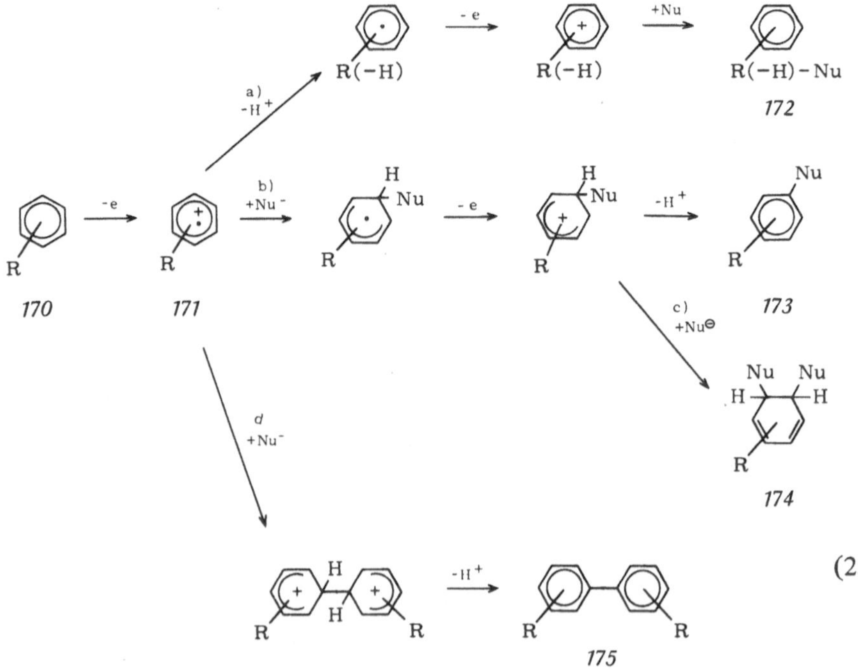

(236)

and deprotonation results in aromatic nuclear substitution (*173*, path b), see 9.1. Addition products *174* are obtained if *171* reacts in a C_NEC_N sequence instead (path c), see 10.1. Dimerization of *171* with subsequent deprotonation yields dimers *175* (path d), 12.1).

The oxidation potentials: $170 \xrightarrow{-e} 171$ of a large number of aromatic hydrocarbons, amines, phenols,heterocycles and olefins are tabulated [10, 10a, 25, 48, 65, 525-528] and need not be repeated here. Such potentials have been successfully correlated with HMO-parameters [525-530, 538]: *i.e.*, in oxidations with the energy of the highest filled MO (HFMO).Adams [25] and Peover [65] have discussed some precaution to which attention should be paid in such correlations, *e.g.*, shifts in potentials due to the irreversibility of the electrode process or due to fast follow-up reactions.

The stability of the radical ion can be judged from electrochemical criteria: the ratio $i_{p,c}/i_{p,a}$ in cyclic voltammetry being close to unity, the constancy of the quantity $i_p/V^{1/2}C$ in single sweep voltammetry or the constant ratio $i_L/\omega^{1/2}C$ at the rotating disc electrode [25]. A further positive proof is the observance of an ESR spectrum. Reviews on the application of ESR to the study of properties and reactions of radical ions are available by Adams [40], Kastening [565e] and Cauquis [566b]. Radical ions can be generated either by electrolysis in the cavity of the ESR spectrometer: *"in situ"* technique = IG, or by external generation = EG by combination of a large area electrode and a flow system. The IG-method allows detection of radicals with life times as short as 10^{-2} to 10^{-3} sec, whereas with the EG technique such with half lives $> 10^{-1}$ sec can be observed [567c]. Application of ESR to electrochemistry in combination with cyclic voltammetry allows direct information on the nature of the radical intermediate. Application of electrochemistry to ESR offers a wider choice of solvents, tolerates impurities and excludes counterion effects on the ESR spectrum.

Generally radical cations are fairly stabile if:

a)   Nucleophilic attack is rendered difficult by a low positive charge density at each atom of the radical cation due to an even charge delocalization to many positions. This is illustrated by the increasing stability of radical cations: anthracene $<$ phenanthrene $<$ tetracene $<$ perylene [25];

b)   Reactive sites are blocked or stabilized by substituents. As an example, while the phenanthrene radical cation is very unstable, its 9,10-diphenyl derivative is long lived.

c)   Polar solvents of low nucleophilicity are used. As an example, radical cations are increasingly stabilized by the solvents acetonitrile $<$ methylene chloride $<$ nitrobenzene.

In Table 17 representative examples of anodically generated radical cations are compiled.

Table 17. *Generation and preparation of radical cations*

Radical cation of	Solvent [a]	$i_{p,c}/i_{p,a}$ [25]	Result	Ref.
Aromatic hydrocarbons:				
9,10-Diphenyl-anthracene	A	1.0 (0,3 V/sec)	Stable for days	531)
9,10-Diphenyl-anthracene	B	1.01 (5 V/min)	ESR (i.G.)	532)
9,10-Diphenyl-anthracene	C		$\tau \sim 50$ min, oxidizes I⁻	533,118)
Rubrene	A	1.0 (0,3 V/sec)	Stable for days	531)
Rubrene	B	1.0 (5 V/min)	ESR	532)
1,3,6,8-Tetraphenyl-pyrene	A	1.0 (0,3 V/sec)	ESR	531)
1,3,6,8-Tetraphenyl-pyrene	B	0.89 (5 V/min)	ESR	532)
Tetracene	A	irrev.		531)
Perylene	B	1.02 (5 V/min)	ESR	532)
9,10-Dimethyl-anthracene	B	0.84 (5 V/min)	ESR	532)
9-Phenylanthracene	B	0.45 (5 V/min)	ESR	532)
Anthracene	C	0.2 (500 V/sec)	Lifetime: few msec	533)
Pyrene	C	0.2 (500 V/sec)	Lifetime: few msec	533)
Chrysene	C	0.2 (500 V/sec)	Lifetime: few msec	533)
Phenanthrene	C	0.2 (500 V/sec)	Lifetime: few msec	533)
Triphenylene	C	Irrev.		533)

Heterosubstituted aromatics:

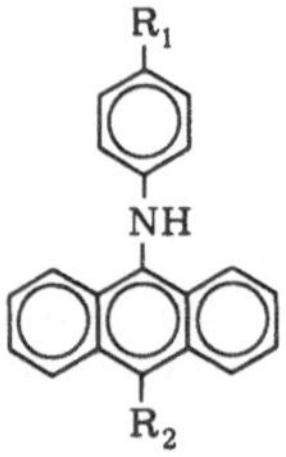 9-Arylamino-anthracene	C	ESR spectra for: R_2=Ph, R_1=OCH$_3$, N(CH$_3$)$_2$	535)
9-Amino-10-phenyl-anthracene	C	ESR	536)
2,4,6-Tri-*t*-butylnitroso-benzene	D	ESR	537)
Hexamethoxybenzene	C	ESR, lifetime $\sim$1 sec, HMO correlation with $E_{1/2}$	538)

Table 17 (continued)

Radical cation of	Solvent[a] $i_{p,c}/i_{p,a}$[25]	Result	Ref.
Pentamethoxybenzene	C	ESR, lifetime ~10 sec, HMO correlation with $E_{1/2}$	538)
1,4-Dimethoxynaphthalene	E	Lifetime 0,2 sec	539)
1,4-Bis (methylthio)-naphthalene	E	Lifetime 0,2 sec	539)
Naphthalene-1,8-disulfide	C	ESR, τ ~5 sec	540)
1,4-Bis (methylthio)-benzene	C	ESR	541)
2,6-Bis (dimethylamino)-naphthalene	E	Lifetime: 15 sec	539)
1,6-Bis (dimethylamino)-pyrene	E	Lifetime: 15 sec	539)
$N(C_6H_5R)_3$ R:pH, p-OCH$_3$, p-CH$_3$, p-Cl, p-Br	C	Very stable, $i_{p,c}/i_{p,a}$ = 1,0, ESR	542)
Pentaphenylpyrrol R: H, CH$_3$, OCH$_3$		UV-data, unresolved ESR	543)
N,N,N',N'-Tetraphenyl-benzidine	E	Lifetime: ca. 15 sec, ESR	539,546)
N,N,N',N'-Tetraphenyl-benzidine	C	Lifetime: ca. 15 sec, ESR	380)
Benzidine		ESR	546)
p-Phenylenediamine	C,F	ESR	544,545)

Heterocycles:

5,10-Dihydro-5,10-di-methylphenazine	C	2 rev. 1e-oxidations (Cyclic voltammetry, chronopotentiometry), ESR	547)
Phenothiazine	C	ESR	549)
1,3,4,7-Tetraphenyl-isobenzofuran	E	τ ~1 sec	548,553)
N-Methyl-1,3,4,7-tetra-phenylisoindole	E	τ > 15 sec	548,553)

Olefins:

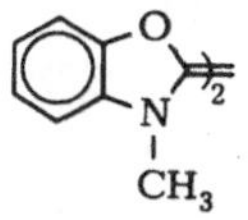

Stability constant: $2 \cdot 10^6$ (by polarography)   550)

Table 17 (continued)

Radical cation of	Solvent[a] $i_{p,c}/i_{p,a}$ [25]	Result	Ref.
		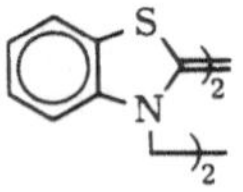 Stability constant: $3 \cdot 10^8$	550)
Tetrakis (dimethyl-amino)-ethylene	C	ESR, $\tau > 1$ hr	386,564)
1,1,4,4-Tetrakis (dime-thylamino) butadiene	C	ESR, $\tau > 1$ hr	386)
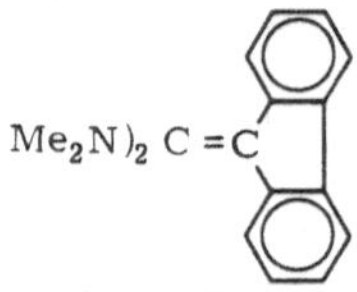 $Me_2N)_2 C = C$	C	ESR, $\tau > 1$ hr	386)

a) A: CH_2Cl_2-$Bu_4N^+ClO^-_4$; nitrobenzene/$Pr_4N^+ClO^-_4$; C: CH_3CN; D: CH_3NO_2/$LiClO_4$; E: DMF; F: aqueous solution.

Fleischmann et al. [534] report cyclic voltammetry data for the oxidation of a series of aromatic hydrocarbons in a molten salt electrolyte, $AlCl_3$-NaCl-KCl at 150°. Electrooxidation in this medium occurs at unusually low oxidation potentials. Tris-(p-substituted phenyl)amines, with the exception of tri (p-nitrophenyl) amine, yield very stable radical cations by all electrochemical criteria [380,542]. Mono- and bis-p-substituted triphenylamines, however, dimerize with rate constants ranging from 10^1 to 10^5 M⁻¹ sec⁻¹ to benzidines 176 (Eq. (237)), which subsequently are oxidized to the radical cations 177, whose ESR-spectra are observed. Dimerization is fastest with the p-NO_2 and p-CN-derivative, in accordance with HMO calculations, which predict the highest spin sensity in the p-position of these compounds [542].

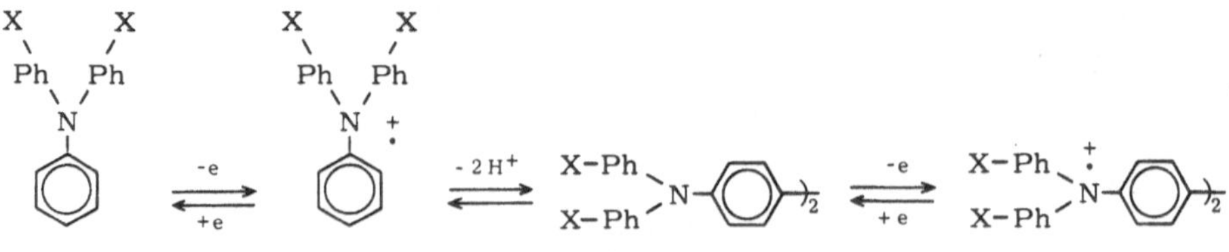

(237)

p-Phenylenediamine *178* is oxidized in a reversible 2e transfer, although simultaneously the ESR spectrum of the radical cation *180* is observed [544]. This result was rationalized by assuming a symproportionation (Eq. (238)) following electron transfer.

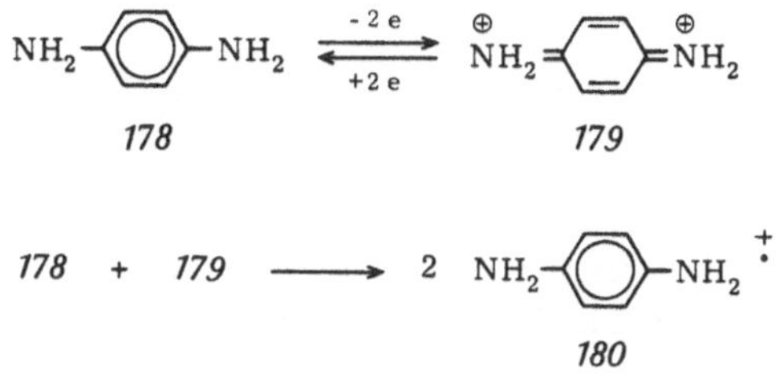

$$178 + 179 \longrightarrow 2 \quad NH_2-\langle\bigcirc\rangle-NH_2 \overset{+}{\cdot}$$

180

(238)

Radical cations of heterosubstituted olefins were obtained by polarographic reduction of the corresponding dications *183* according to the scheme in Eq. (239) [550]:

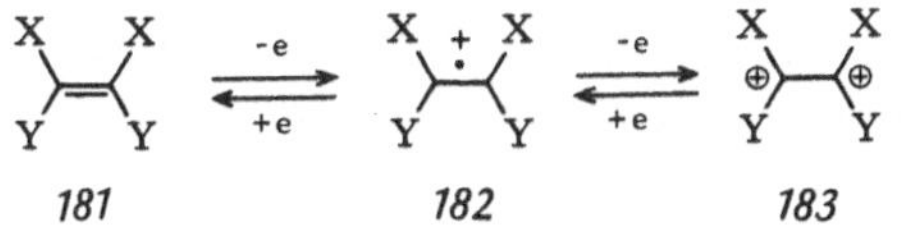

181   *182*   *183*

(239)

The stability constant $K = \dfrac{[182]^2}{[181] \cdot [183]}$ for various radical cations was determined from the index potential $E_{182/183}$ and $E_{181/182}$ as to be $K = 3 \cdot 10^8$ for *184*, $K = 2 \cdot 10^6$ for *185*, $K = 7 \cdot 10^5$ for *186* and $K = 20$ for *187*.

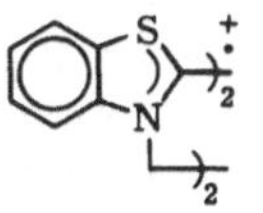

184 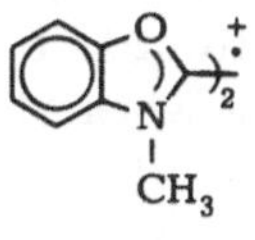   *185* 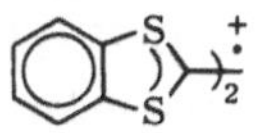   *186* 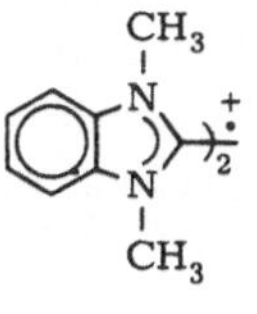   *187*

Fritsch and Weingarten [386] have studied the electrooxidation of 22 substituted enamines by cyclic voltammetry and ESR spectroscopy. Oxidation potentials of these strong electron donors may be as low as -0,901 V (*vs.*S.C.E.). The life times of the initially formed radical cations range from 0,005 sec to days depending on the efficiency of reactive site blocking or stabilization by substituents. Coupling constants from the ESR spectra indicate that the unpaired electron is polarized away from the dimethylamino substituents. The opposite is true for the

145

positive charge, so that the charge distribution in the radical cation is best represented by *188*.

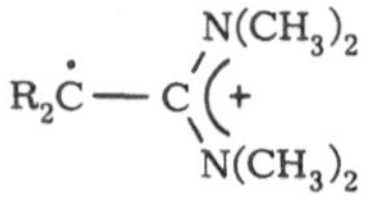

$$R_2\overset{.}{C} - C \Big(\!\!\begin{array}{l} N(CH_3)_2 \\[4pt] + \\[4pt] N(CH_3)_2 \end{array}$$

188

14.2. Electrochemiluminescence

Academic and commercial interest in electrochemiluminescence = ECL gave a great impetus to research in elechtrochemically generated radical ions. The simplified scheme for ECL is shown by Eq. (240):

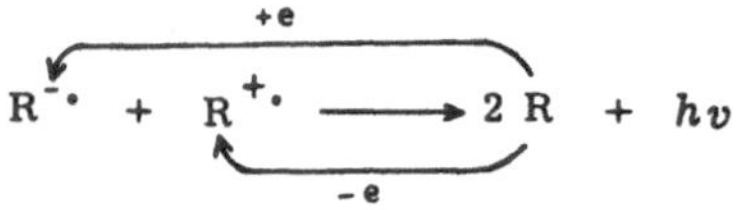

$$\overset{+e}{\overbrace{R^{-\cdot} + R^{+\cdot}}} \longrightarrow 2\,R + h\upsilon \tag{240}$$
$$\underbrace{\phantom{R^{-\cdot} + R^{+\cdot}}}_{-e}$$

By annihilation of a radical cation and a radical anion neutral species are formed and light is emitted. Reduction and oxidation of the neutral compounds regenerates the radical ions for a new cycle. By fast repetition of this cycle, *e.g.*, *via* electrolysis on an AC-line, continous conversion of current into light should be possible.

ECL has been reviewed by Hercules [551] and Kuwana [552]. Zweig *et al.* [539, 553] have investigated the use of a great variety of polycyclic aromatics for ECL, *e.g.*, 1,4-dimethoxynaphthalene, 1,6-bis(methylthio)pyrene or 1,3,4,7-tetraphenyl-isobenzofuran. According to their results the combined requirements for successful continous ECL in cyclic processes have to be high radical anion and cation stability, high fluorescence efficiency, high solubility of the compounds and chemical stability of the neutral hydrocarbons. Mostly the first condition cannot be met, since either the radical anion or the radical cation is fairly unstable and is consumed in unwanted chemical decays, which end ECL after some cycles. Hercules [554] initiated the studies on ECL by subjecting aromatic hydrocarbons to AC-electrolysis in CH_3CN or DMF/TEAB. Strong ECL was found for pyrene, rubrene, perylene or 9,10-diphenylanthracene when AC of 100 cps was used for electrolysis [555]. The emission spectrum was, as in most other examples, identical with the fluorescense spectrum of the hydrocarbon. This is rationalized by the annihilation mechanism shown below (Eq. 241)):

$$R^{+\cdot} + R^{-\cdot} \longrightarrow R^{S_0} + R^{S_1} \overset{h\upsilon}{\longrightarrow} 2\,R^{S_0}$$

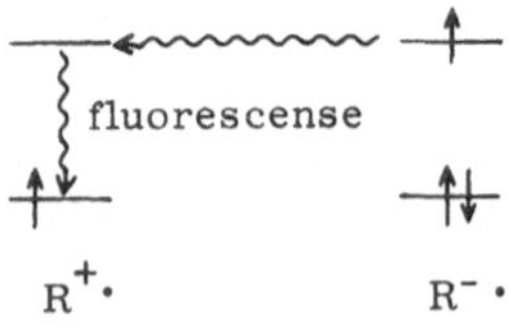

$$\tag{241}$$

The electron from the antibonding orbital of the radical anion is transferred to the antibonding orbital of the radical cation. The $\pi^* \to \pi$ transition of this electron to the ground state produces the fluorescence spectrum.

Bard *et al.* [556,557)] and Visco *et al.* [558)] have quantitatively analyzed the intensity of pulsed ECL of 9,10-diphenylanthracene, tetraphenylpyrene and rubrene. By computer simulation of the electrode process and the subsequent chemical reactions the rates for chemical decay of the radical ions could be determined. Weaker ECL with fluorescence emission [559)] or electrophosphorescence [560)] occurs if the radical anion $R^{-\cdot}$ reacts with a dissimilar radical cation $R'^{+\cdot}$ of insufficient high oxidation potential to gain enough energy for fluorescence emission, that is, if $h\nu$(fluorescence) > 23.06 ($E_{R'+}. -E_{R^-}.$), *e.g.*, in the annihilation of the anthracene radical anion with Wurster's blue. For these process the following schemes are assumed (Eq. (242)):

$$R_1^{+\cdot} + R_2^{-\cdot} \longrightarrow R_1 + R_2^T; \quad R_2^T \longrightarrow R_2^{S0} \quad \text{electrophosphorescene}$$

$$R_2^T + R_2^T \longrightarrow R_2^{S0} + R_2^{S1}; \quad R_2^{S1} \longrightarrow R_2^{S0} \quad \text{weaker fluorescence} \quad (242)$$

The triplet emission can be totally suppressed by adding an efficient quencher, *e.g.*, 1,3,5-hexatriene [560)].

14.3. Anodic Generation of Radicals

2,4,6-Tris (*t*-butyl) phenoxy-, 4,4'-methylenebis (2,4- di-*t*-butyl) phenoxy- or 4,4'-thiobis (2-*t*-butyl-6-methyl) phenoxy-radicals have been prepared for ESR observation by oxidation of the corresponding phenolates at a graphite electrode in acetonitrile [561)]. Dimroth *et al.* [562)] determined reversible 1e -oxidation potentials for a series of substituted phenols in basic medium. $(CF_3)_2NOH$ has been oxidized in alkaline medium on platinum or magnetite in quantitative yield to the pink violet hexafluorodimethylnitroxide $(CF_3)_2NO \cdot$ [563)].

14.4. Generation of Radical Anions

Radical anions *189* (Eq. (243)) are generated by electron transfer from the cathode or a reducing agent to the lowest empty orbital (LUMO) of a neutral substrate. The electrochemical generation of the radical anion (or cation) is superior to chemical reduction (or oxidation). Advantages are the use of a wider range of solvents, the applicability of an electrode potential that can be regulated at will for selective radical ion formation, and a greater choice of counterions, which greatly influence the properties of radical ions. As there are extensive and authoritative reviews on radical anions available [111a,565,569] a short summary on radical anion reactions is sufficient at this place.

Reduction potentials of miscellaneous compounds have been determined and are tabulated elsewhere [65,111a,566]. Good correlations of reduction potentials of aromatics with the energies of the corresponding LUMO's have been obtained [567]. A detailed and comprehensive review on the polarographic behavior of aromatic hydrocarbons is given by Peover [65,568]. Most reactions of the radical anions *189* are essentially the mirror image reactions of radical cations. The radical anion *189* can dissociate to a radical and an anion Nu⁻ (Eq. (243), path a), with the radical being subsequently reduced and protonated to the cathodic cleavage product *190* (see 13.2). This sequence corresponds to path a) and b) for radical cation decay in Eq. (236). Protonation of *189* followed by reduction and a second protonation results in formation of addition product *191* (path b) (see 10.2 on cathodic addition). This path is equivalent to c) in Eq. (236). Coupling of *189* with subsequent protonation yields the dimer *192* (path c, see 12.2 on cathodic coupling), a path that corresponds to d) in Eq. (236). Furthermore, *189* may disproportionate (path d), act as a reducing agent by electron transfer (path e), or add to activated double bonds (path f) to produce the dimer *194* (see 12.2.).

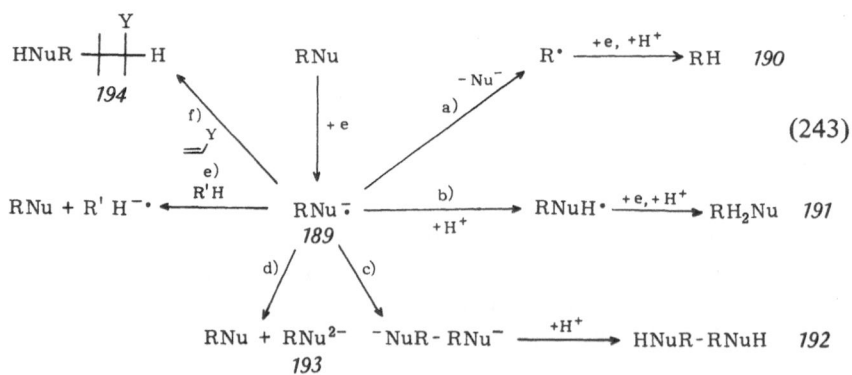

$$(243)$$

Protonation of *189* or the dianion *193* occurs at the sites of highest negative charge density, which however may be modulated by the electric field strength at the electrode/electrolyte interface *via* the polarity of the solvent [571].

The dimerization of two radical anions *189* to the dianion *195* (Eq. (244)) depends on a) how effectively the counterion ME⁺ compensates the negative charges in *195* to reduce the electrostatic repulsion,   b) the gain in energy by

formation of the new C-C bond and   c) the loss of resonance energy in *189* due
to bond formation. As examples, the radical anions of styrene, α-methylstyrene
and 1,1-diphenylethylene dimerize exclusively, whereas those of tri- or tetraphenylethylene exist as monomers. The dimerization rates k_f for radical anions
are of the order of 10^4 to 10^6 M^{-1} sec^{-1}, [111a)] which is significantly slower than
for radical dimerizations ($k \sim 10^8$ M^{-1} sec^{-1}) [566a)]. The dissociation rate k_b was
determined for the α-methylstyrene dimer dianion to be 10^{-8} sec^{-1}. [570)]

$$2\,ME^+R^{-\cdot} \quad \underset{k_b}{\overset{k_f}{\rightleftharpoons}} \quad ME^{+\,-}R\text{-}R^-\,ME^+ \qquad\qquad (244)$$

$$\phantom{2\,ME^+R^{-\cdot}}\;189\phantom{\underset{k_b}{\overset{k_f}{\rightleftharpoons}}}\qquad\qquad 195$$

The disproportionation equilibrium (Eq. (245)) is shifted the more towards
the side of *193* and *196*, the better the negative charges in *193* are compensated
by the counterion. This is achieved by the use of small counterions, *e.g.*, K_D
increases with Cs $<$ K $<$ Na $<$ Li, or electrolytes which poorly solvate cations,
e.g., K_D *increases* with the solvent in the order HMPA $<$ glyme $<$ THF $<$ diethyl
ether $<$ dioxane. Some reductions to dianions, appearing as 2e-transfers, are actually 1e-reductions to radical anions which subsequently disproportionate, *e.g.*,

$$2\,ME^+R^{-\cdot} \quad \overset{K_D}{\rightleftharpoons} \quad R + ME^{+\,-}R^{-\,+}ME \qquad\qquad (245)$$

$$\phantom{2\,ME^+R^{-\cdot}}\qquad\qquad 196\qquad 193$$

the reductions of tetraphenylethylene (Eq. (246)), benzophenone, or anthraquinone to the corresponding dianions [572)].

$$(C_6H_5)_2C{=}C(C_6H_5)_2 \xrightarrow{+e} (C_6H_5)_2\,\bar{C}\text{-}\dot{C}(C_6H_5)_2 \qquad (C_6H_5)_2\bar{C}\text{-}(\bar{C}_6H_5)_2 \quad (246)$$

Radical anions equilibrate with excess neutral precursors according to
Eq. (247):

$$ME^+\,R_a^{-\cdot} + R_b \overset{k_t}{\rightleftharpoons} \quad R_a + ME^+\,R_b^{-\cdot} \qquad\qquad (247)$$

The electron transfer rates k_t range in the order of 10^7 to $10^8 M^{-1} sec^{-1}$. Furthermore the radical anion may transfer its electron to each suitable acceptor which affords an energetically more favorable radical anion. As an example, sodium biphenylene reacts with anthracene to form the anthracene radical anion exclusively [111a]. Polymerizations by radical anions may be preceded by an electron transfer to the monomer (Eq. (248)). Thus the monomer radical anion may initiate anionic and/or mixed anionic-radical polymerization.

$$R^- \cdot + M \longrightarrow R + M \cdot^- \begin{array}{c} \nearrow ^-M\text{-}M^- \\ \downarrow M \\ \searrow \cdot M\text{-}M^- \end{array} \tag{248}$$

Reactions of radical anions with halides are believed to occur *via* electron transfer and subsequent radical coupling according to Eq. (249) [565].

$$R'\text{-}Y + R^{-} \cdot \longrightarrow R'Y^{-} \cdot + R \longrightarrow R'' + Y^- \xrightarrow{R^{-} \cdot} R'\text{-}R^- \tag{249}$$

$$R'\text{-}R^- \xrightarrow{R} R'\text{-}R \cdot + R^{-} \cdot$$

Some representative examples of cathodically generated radical anions are compiled below (Table 18).

Reduction of cyclooctatetraene (COT) in DMF or DMSO was studied by cyclic voltammetry, DC and AC polarography, and ESR spectroscopy [574]. COT is reduced in aprotic media in two 1e-transfers (Eq. (250)). Cpe at -1,8 V in the

$$COT \xrightarrow{-1,62 \text{ V}} COT^- \cdot \xrightarrow{-1,86 \text{ V}} COT^{--} \tag{250}$$

ESR cavity produces a signal (nonuplet, $a_H = 3,23$ G), which indicates eight equivalent protons. When bicyclo [6.1.0] nonatriene (*196*) was reduced in the ESR cavity the spectra of the cyclononatriene radical anion (*197*) and of the methylcyclooctatetraene radical anion (*198*; Eq. (251)) were observed.

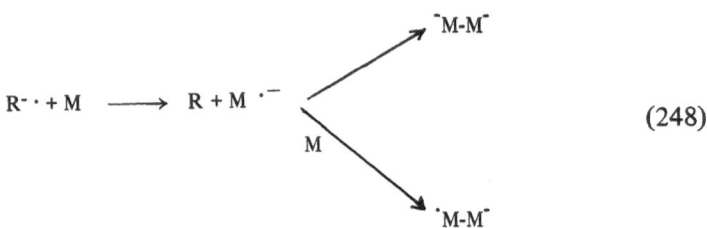

$$\tag{251}$$

196   *197*   *198*

Benzocyclobutadienequinone (*199*)[578], fluorenone, benzil, 4,5-phenanthryl-eneketone (*200*), *o*-phenanthrenequinone and acenaphthenequinone[579] yield stable ketyls on cathodic reduction in DMF or acetonitrile, and these have been characterized by ESR. An interesting electrocyclic isomerization of the anion radical *201* (Eq. (252)) to the 2,2'-bisallyl radical anion *202* was observed by ESR, when 1,2-dimethylenecyclobutane (*203*) was reduced in THF at -90°C[581].

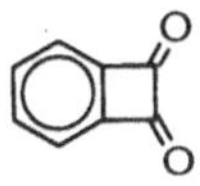

199 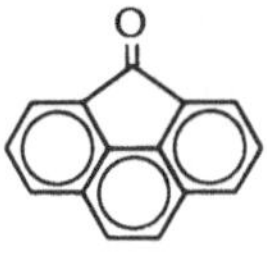   *200*

(252)

203   *201*   *202*

Maki and Geske[582,583] have reported detailed ESR studies of radical anions derived from a series of substituted nitrobenzenes. These were generated by "in situ" electrolysis in acetonitrile / $Pr_4N^+ClO_4^-$. The spin densities calculated from the coupling constants were in qualitative agreement with HMO-predictions[582]. For substituted nitrobenzenes a linear correlation between the reduction potential and the a_N-coupling constant was established. The ESR spectra of primary, secondary, and tertiary nitroaliphatic radical anions have been obtained and analyzed[524,590]. The lifetime of these species is mostly less than 1 sec. As expected the spin density at nitrogen, $a_N = 25G$, is much higher than at the nitrogen of nitrobenzene radical anions, $a_N = 10$ G. Aliphatic nitro radical anions decay to long-lived dialkylnitroxides. Hoffmann *et al.*[524] have made a detailed study of this decomposition and rationalized the products by the following scheme (Eq. (253)) (see also 13.2):

$$RNO_2 \xrightarrow{+e} RNO_2^{-\cdot} \longrightarrow R^\cdot + NO_2^-$$

$$R^\cdot \longrightarrow R\text{-}R, RH, RH(\text{-}H_2)$$

$$RNO_2^{-\cdot} + R^\cdot \longrightarrow RNO\text{-}OR^- \longrightarrow R\text{-}N=O + OR^-$$

$$R\text{-}N=O + R^\cdot \longrightarrow R_2N=O^\cdot$$

(253)

Fraenkel *et al.*[591] have determined ESR coupling parameters and spin densities of various substituted aromatic and aliphatic nitrile radical anions. These radical anions react in suitable cases by elimination of cyanide ion (Eq. (254))

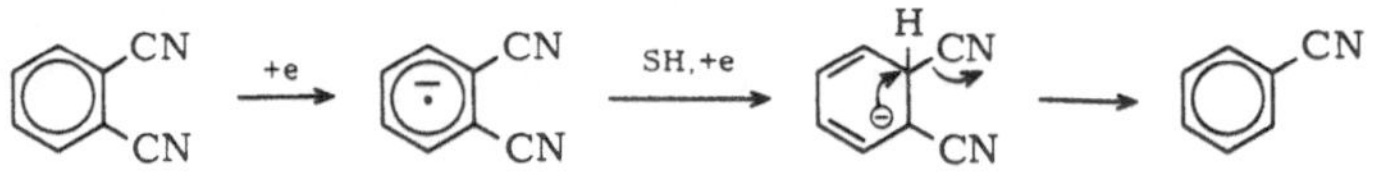

$$(254)$$

or dimerize with subsequent elimination of substitutents, *e.g.*, F, NH$_2$ (Eq. (255)).

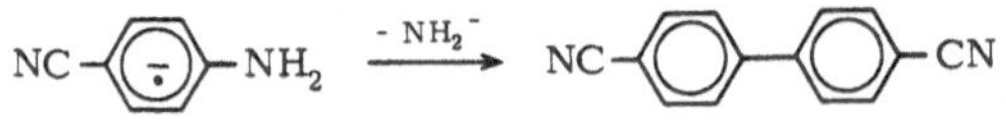

$$(255)$$

Cathodic reduction of 1,1,2,2-tetracyanocyclopropane or 1,1,2,2,-tetracyano-ethane yields the radical anion of tetracyanoethylene *via* a formal reductive de-methylenation and dehydrogenation, respectively (Eq. (256)) [592].

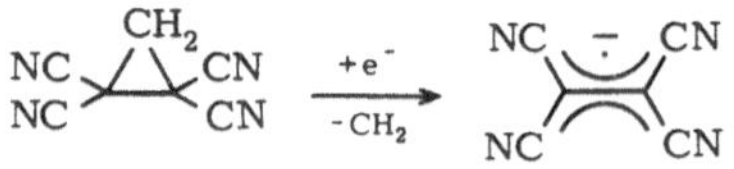

$$256)$$

Dessy *et al.* [593] generated the mercury substituted radical anion of nitro-benzene for ESR observation by reduction of bis (*p*-nitrophenyl)mercury. The a_{Hg} coupling constant was found to be 10 times larger than the corresponding aH coupling constant, an amplification which is useful for the resolution of small a_H coupling constants by substitution of H *versus* Hg.

Radicals, probably radical anions, were detected in the cathodic cleavage of benzoyl-N,N-diphenylamine, benzoyl-N-phenyl-N-methylamine, triphenylphos-phinoxide, and others [594]. No signals were obtained in the electrolysis of phenyl benzoate or phenyl sulfinate.

Table 18 gives some further examples of cathodically generated radical anions.

Table 18. *Preparation and properties of radical anions*

Radical anion of	Solvent[a]	Results	Ref.
Aromatic hydrocarbons:			
Benzene	C	ESR (septett, aH=3,5 G, -70°C)	291)
9,10-Diphenylanthra-cene		Polarography, chronopotentiometry, coulometry, cyclic voltammetry $i_{p,c}/i_{p,a} = 1$	573)
Anthracene		ESR, polarography	572)

152

Table 18 (continued)

Radical anion of	Solvent[a]	Results	Ref.
Heterosubstituted aromatics:			
9-Methoxyanthracene	A	Lifetime 15 sec	539)
10,10'-Dimethoxy-9,9'-bisanthranyl	A	Lifetime 15 sec	539)
2-Dimethylamino-naphthalene	A	$\tau \sim 15$ sec	539)
NO_2—⬡—X—⬡—NO_2		ESR	584)
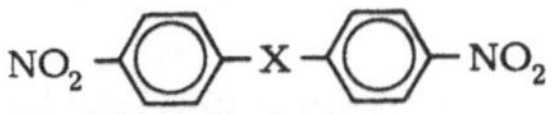 $X=CH_2, O, S, CH_2\text{-}CH_2$			
o-, *m*-Trifluoromethyl-nitrobenzene		Cyclic voltammetry, polarography	589)
p-Trimethylsilylnitro-benzene		ESR	585)
Trinitromesitylene		ESR	586)
Dinitrodurene		ESR	586)
Dinitrobenzene	A	ESR	587)
Nitronaphthalene	A	Visible spectra	588)
Nitromesitylene	A	Visible spectra	588)
Benzophenone		ESR, polarography	572)
Anthraquinone		ESR, polarography	572)
Olefins:			
Butadiene	B	ESR, $\tau_{-115°} = 30$ Min	576)
2,5-Dimethyl-2,4-hexa-diene	B	ESR, $\tau_{-80°} = 15$ sec	576)
Isoprene	D	ESR (-90°)	577)
1,3-Cyclohexadiene	D	ESR (-90°)	577)
2,3-Dimethylbutadiene	D	ESR (-90°)	577)
Tolane	B, C	ESR	580)
Stilbene	A, B, C	ESR, $\tau \sim 10$ sec (A)	443,580)
Azobenzene	B, C	ESR, $\tau \sim 10$ sec (A)	580)
Dibenzoylethylene	A	$\tau \sim 15$ sec (25 °C)	443)
2,2,6,6-Tetramethyl-4-heptene-3	A	ESR	418)
Heterocyclics:			
1,3,4,7-Tetraphenyl-isobenzofuran	A	$\tau \sim 15$ sec	553)
Phenazine	B	ESR, $\tau_{220} = 2-4$ min	601,556)
Phthalimide	A	ESR	595)
Naphthalimide	A	ESR	595)
Phenazine-N, N-dioxide	A	ESR	596)

a) A: DMF, B: THF, C: glyme, D: NH_3

14.5. Cathodic Generation of Radicals

Radicals can be generated by reduction of carbonium ions or onium salts. Reduction of the N-alkylpyridinium salt *203* yields in either buffered aqueous KCl or $CH_3CN/Bu_4N^+ClO_4^-$ the radical *204* in 100% efficiency [597].

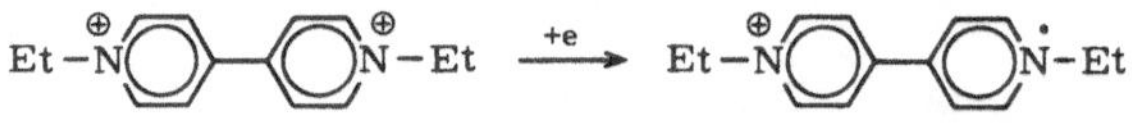

203   *204*

Reduction of substituted triarylcarboniumions to the corresponding radicals was studied by polarography [598] and cyclic voltammetry [599]. The half wave potential is shifted to more negative values with increasing stability of the carbonium ion by *p*-substituents R in the order R:

Cl > F > phenyl > *t*-butyl > CH_3 > cyclopropyl > OCH_3 > $N(C_2H_5)_2$ [598].

The reduction of $(C_6H_5)_3C^+$ $SbCl_6^-$ occurs at +0,46V (versus *Ag/AgCl*) in a reversible 1e-process, whilst the subsequent reduction of the trityl radical at -1,07 V is irreversible [599]. Reduction potentials for diarylmethyl cations, which were generated by protonation of the corresponding alcohols or olefins by 97% sulfuric acid or by anhydrous $HClO_4$ in methylene chloride have been reported [600].

15. Indirect Electrochemical Process

In the preceding sections, we have been mainly concerned with *direct* electrochemical reactions, *i.e.*, reactions in which electron transfer takes place directly
between the electrode and the organic substrate. In an *indirect* process, an added
reagent — most commonly an inorganic ion of some kind — is converted electrochemically to a species which then attacks the substrate in a homogeneous reaction with formation of the desired product and, most commonly, regeneration of
the added reagent. Here the electrolytic procedure is merely used as a convenient
source for otherwise difficultly accessible reagents and the preparative result does
not differ from the corresponding homogeneous reaction. In effect, many of these
reactions are "catalytic" in the added reagent since only substrate, electric energy,
and solvent are consumed in the over-all reaction. There are also cases known
where a truly catalytic reagent can be formed electrochemically.

Several types of intermediates can be generated electrochemically for use in
an indirect process. These include:

1. Metal ions in unusual valence states,
2. Amalgams,
3. Nucleophilic and basic reagents,
4. Halogens,
5. Organic intermediates.

The mode of operating indirect electrolytic reactions can be either the conventional one, *i.e.*, by having the substrate as a component of the electrolyte, or
be a two-step procedure in which the electrolyte containing the species generated
is allowed to react with the substrate in a separate vessel, preferably by a continuously operating system.

15.1. Generation of Metal Ions

Manganic sulfate can be generated anodically by electrolysis of manganous sulfate
in 55% sulfuric acid at a lead anode [157]. After electrolysis, xylene is added and
allowed to react at 25-35°, tolualdehyde being formed in 30-45 % current yield.
The used electrolyte is then purified and used again.

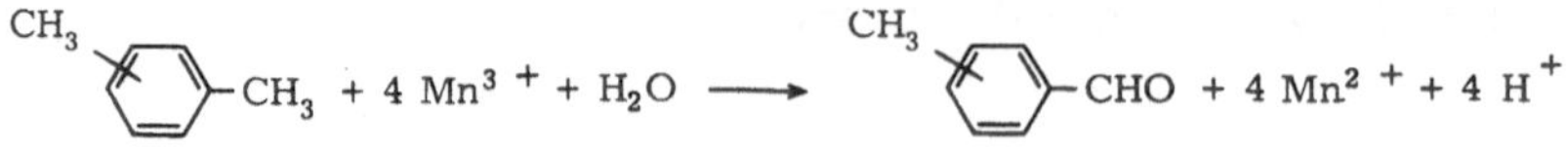

$$CH_3-\text{(ring)}-CH_3 + 4\ Mn^{3+} + H_2O \longrightarrow CH_3-\text{(ring)}-CHO + 4\ Mn^{2+} + 4\ H^+$$

$$(257)$$

Similarly, cobaltic and argentic ion have been generated by anodic oxidation of cobaltous and argentous ion, respectively, and used for the oxidation of methyl-substituted aromatic hydrocarbons to aldehydes [18a]. Electrogenerated mercuric ion can be used for the conversion of propene to acrolein [18a]:

$$CH_3CH=CH_2 + 4\ Hg^{2+} + H_2O \longrightarrow CH_2=CH\text{-}CHO + 4\ H^+ + 2\ Hg_2^{2+} \quad (258)$$

$$Hg_2^{2+} \longrightarrow 2\ Hg^{2+} + 2\ e^-$$

An unusual type of reaction is "anodic reduction" which can be performed at certain metal anodes. Thus, when magnesium is used as an anode in the electrolysis of benzophenone in pyridine/sodium iodide solution, the anode is consumed and benzopinacol can be isolated from the anolyte [158]. Here reduction by univalent magnesium ion is postulated:

$$Mg \longrightarrow Mg^+ + e^-$$

$$2\ Mg^+ + 2\ Ph_2C=O \longrightarrow 2\ Mg^{2+} + \begin{array}{c} Ph_2C\text{-}O^- \\ | \\ Ph_2C\text{-}O^- \end{array} \quad (259)$$

Similarly, an aluminium anode is anodically converted to Al^+ which for example can reduce nitrosobenzene in the anode compartment [159]:

$$Al \longrightarrow Al^+ + e^-$$

$$(260)$$

$$3\ Al^+ + 2\ PhNO + 4\ H_2O \longrightarrow 6\ OH^- + 3\ Al^{3+} + PhNHNHPh$$

"Cathodic oxidations" also become possible in certain cases. Hydroxylation of aromatic hydrocarbons can be achieved by oxidation with oxygen in the presence of some reduced metal ions, e.g., Cu(I) [160]. The method has been adapted for electrochemical regeneration of Cu(I) from Cu(II) in acetic acid solution [161]; a 36 % current yield (76 % yield based on hydrocarbon consumed) of cresols plus cresyl acetates could be realized in the case of toluene:

$$Cu^{2+} + e^- \rightarrow Cu^+$$

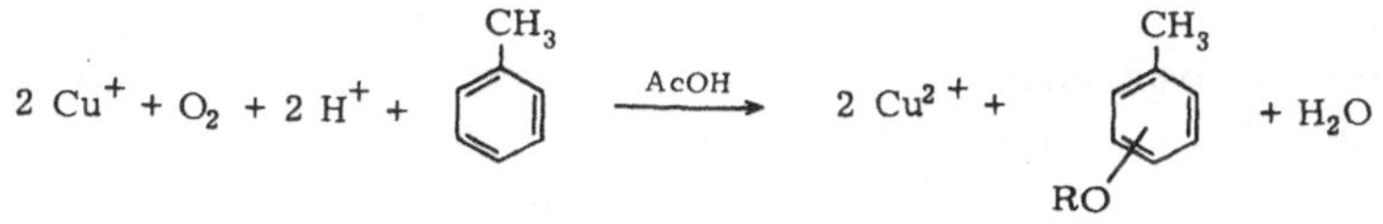

(261)

$$2\,Cu^+ + O_2 + 2\,H^+ + \text{[toluene]} \xrightarrow{AcOH} 2\,Cu^{2+} + \text{[substituted toluene, RO]} + H_2O$$

(R=H or Ac)

15.2. Generation of Amalgams

The use of electrogenerated sodium amalgam for the hydrodimerization of activated olefins (cf. Sect. 12.2) has been studied in some detail [162,163]. Quaternary ammonium amalgams have been prepared by electrolysis of quaternary ammonium salts in acetonitrile or dimethylformamide with rigorous exclusion of oxygen [164]. These amalgams react with acrylonitrile in acetonitrile-water mixtures to give 100 % adiponitrile with only trace amounts of propionitrile being formed. Reaction with butyl iodide produced 25 % dibutylmercury.

15.3. Generation of Nucleophiles and Bases

Oxygen is reduced at the mercury cathode in a dipolar aprotic solvent containing a quaternary ammonium salt to form the superoxide ion, $O_2^{-\cdot}$, in a reversible one-electron transfer process [165,166]. The reduction takes place at -0.8 V (SCE) and concentrations of tetraalkylammonium superoxide of at least 0.1 M can be obtained (the half life of the superoxide ion in dimethylformamide is about 40 min at room temperature).

Superoxide ion is a strong base and can be used for initiation of the autoxidation of certain aromatic hydrocarbons. Thus, electrolysis of fluorene at -0.8 V (SCE) in DMF-tetrabutylammonium perchlorate solution continuously flushed with oxygen produces fluorenone with a current efficiency of about 5000 %. The reaction was demonstrated to be a base-catalyzed oxidation, in which superoxide ion acts as a base [166a].

Superoxide ion is also a powerful nucleophile, capable of reacting with alkyl halides to form dialkyl peroxides:

$$R\text{-}X + O_2^{-\cdot} \longrightarrow R\text{-}O_2{}^\cdot + X^-$$

(262)

$$RO_2{}^\cdot + O_2^{-\cdot} \longrightarrow R\text{-}O_2^- + O_2$$

(263)

$$R\text{-}O_2^- + R\text{-}X \longrightarrow R\text{-}O\text{-}O\text{-}R + X^-$$

(264)

From butyl bromide, an 80 % yield of dibutyl peroxide was obtained.

Electrochemical generation of strong base has been applied in an interesting modification of the Wittig reaction [167]. A solution of azobenzene, benzaldehyde, benzyltriphenylphosphonium bromide, and lithium chloride in dimethylformamide was electrolyzed at a potential where only the easily reducible azobenzene was electroactive. The phosphonium ion then acts as a proton source, giving the ylide as an intermediate. Under these conditions a 98 % yield of a mixture of *cis-* and *trans*-stilbene was obtained, presumably *via* the following reactions:

$$PhN=NPh \xrightarrow[-0.9 \text{ V}]{2e, \; 2 \, H^+} PhNH\text{-}NHPh \qquad (265)$$

$$Ph_3\overset{+}{P}CH_2Ph \xrightarrow{-H^+} Ph_3P=CHPh \qquad (266)$$

$$Ph_3P=CHPh \xrightarrow{PhCHO} Ph_3PO + Ph\text{-}CH=CH\text{-}Ph \qquad (267)$$

98%   *cis*:   59%

trans: 39%

15.4. Generation of Halogens (see also Sect. 9.1).

Numerous cases of electrochemical halogenation (chlorination, bromination, iodination) are known and have been discussed in some detail by Allen [6]. Generally, these reactions appear to be indirect electrolytic processes, since in most cases the substrates are oxidized at higher potentials than chloride, bromide, and iodide ion. More work is needed, however, to clarify anodic halogenation mechanism.

Electrochemical fluorination [168,169] is a commercial process for perfluorination of aliphatic compounds. The reaction is performed in liquid hydrogen fluoride -potassium fluoride at a nickel anode. The mechanism is not known; free fluorine cannot be detected during electrolysis, so it seems probable that fluorination is a direct electrochemical reaction. Theoretically, hydrogen fluoride-potassium fluoride should be a very oxidation-resistant SSE, and it might well be that the mechanism is analogous to that proposed for anodic acetamidation of aliphatic compounds in acetonitrile-tetrabutylammonium hexafluorophosphate [44].

There has been a great deal of commercial interest in an electrochemical process for the production of propylene oxide from propene *via* anodic generation of halogen [170,171]. The reactions are summarized below (X = Cl, Br):

Anode process: $\qquad 2\,X^- \longrightarrow X_2 + 2e^-$ (268)

Cathode: $\qquad 2\,Na^+ + 2\,H_2O + 2e^- \longrightarrow H_2 + 2\,NaOH$

(269)

Chemical process near anode: $\qquad CH_3CH{=}CH_2 + X_2 + H_2O \longrightarrow CH_3CH(OH)CH_2Br + HBr$

Chemical process near cathode: $\qquad CH_3CH(OH)CH_2Br + HBr + 2\,NaOH \longrightarrow$ (270)

$$CH_3\underset{\underset{O}{\diagdown\diagup}}{CH}{-}CH_2 + 2\,NaBr + 2\,H_2O$$

Overall reaction: $\qquad CH_3CH{=}CH_2 + H_2O \xrightarrow[\text{energy}]{\text{electrical}} CH_3\underset{\underset{O}{\diagdown\diagup}}{CH}{-}CH_2 + H_2$ (271)

This process is said to compete favorably with the existing commercial procedure.

15.5. Generation of Organic Intermediates

Under this heading a few borderline cases between direct and indirect electrochemical reactions can be included. Dihalocarbenes are probably intermediates in the cathodic reduction of certain polyhalogenated compounds [172]; as an example, cathodic reduction of carbon tetrachloride at a mercury cathode in acetonitrile-tetrabutylammonium bromide in the presence of tetramethylethylene gave a low yield of 1,1-dichloro-2,2,3,3-tetramethylcyclopropane:

$$CCl_4 \longrightarrow CCl_3^- + Cl^-$$ (272)

$$CCl_3^- \longrightarrow\; :CCl_2 + Cl^-$$

$$(CH_3)_2C{=}C(CH_3)_2 +\; :CCl_2 \longrightarrow (CH_3)_2\underset{\underset{CCl_2}{\diagdown\diagup}}{C}{-}C(CH_3)_2$$ (273)

Some evidence for the formation of benzyne in the cathodic reduction of *o*-dibromobenzene [173] has been obtained:

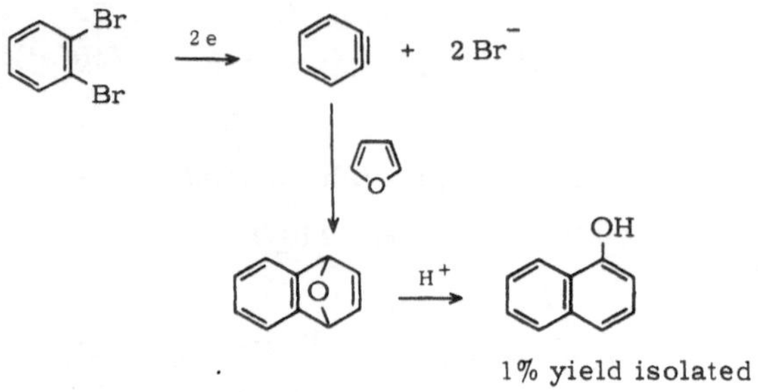
Br
Br
2 e
+ 2 Br⁻
H⁺
OH
1% yield isolated

(274)

Attempted anodic bis-decarboxylation of *o*-phthalic acid (as its dianion) failed to give any evidence for benzyne formation [174].

16. Electropolymerization

Transient electrogenerated species can be used for the initiation of polymerization processes of different kinds [635,636,637]. The main advantage of this type of polymerization appears to be the possibility of controlling the polymerization process by changing the electric current, *i.e.*, the rate of formation of the initiator, which can be a neutral radical, a radical ion, a carbonium ion, or a carbanion. More over, there is - at least in principle - the attractive possibility that chain growth may take place in a stereoregular fashion at the electrode surface. Although in most cases electrochemically initiated polymerization does not appear to be sterically controlled in this manner, a recent report shows that some stereoregularity can indeed be induced in certain cases, *e.g.*, in the cathodic polymerization of phenyl isocyanate [638], and hence holds promise that applications of this principle might be found.

Electrolytically initiated polymerization may either depend on a direct electron transfer between electrode and monomer, or on the formation of an intermediate which interacts with a monomer molecule in a fast chemical step, thus creating a chain initiator. As an example of the former type of process, the formation of a living polymer from the cathodic polymerization of α-methylstyrene by electrolysis in sodium tetraethylaluminate - tetrahydrofuran may be cited [639], whereas a typical case of the latter type is the anodic polymerization of vinyl monomers by electrolyzing them together with sodium acetate in aqueous solution [637,640]. Here it is assumed that acetate ion is discharged to form an acetoxy or methyl radical which attacks the monomer molecule in a fast chemical step. This technique is useful for monomers which are difficult to polymerize by other techniques. Thus, chlorotrifluoroethylene can be polymerized anodically with good efficiency by electrolysis in liquid hydrogen fluoride - potassium fluoride or trifluoroacetic acid - trifluoroacetic anhydride - potassium trifluoroacetate [637].

17. Organometallics

This section deals with some electrolytic preparations of organometallics. A review covering part of the subject is available [610].

Radical ions from transition metal complexes:

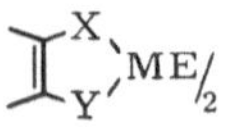

$$\begin{array}{c} X \\ \diagdown \\ \diagup \\ Y \end{array} ME_{\diagup_2}$$

with X, Y = O, NH, S and ME = Ni, Co, Zn[609] or LME $(CO)n$ with L = ligands such as triphenylphosphine, pyridine, alkyne, cyclopentadiene and ME = Cr, Mo, W [623-632], or from boron cage species $B_{10}H_{14}$, $B_{10}H_{12}{}^{2-}$[611] can be generated electrochemically.

In these radical ions the unpaired electron may be polarized to the ligands or localized at the metal with all intermediate situations existing in between. The detailed electron distribution can be resolved by applying a combination of ESR, IR, NMR, and Mössbauer spectroscopy [612].

17.1. Anodic Generation of Organometallics

Organometallics can be generated in laboratory or industrial scale by anodic transmetalation: $RME_1 \longrightarrow RME_2$ (Eq. (275)).

$$RME_1 \xrightarrow{\;-e\;} R^\cdot + ME_1^+$$

$$R^\cdot + ME_2 \longrightarrow RME_2 \qquad\qquad (275)$$

It is assumed that the organometallic, RME_1, is oxidized to metal ion ME_1^+ and radical R which reacts with active anodes ME_2 to form RME_2. An S_E2-reaction of ME_2^+, formed at the anode surface, with RME_1 may however also apply for some transmetalations. Table 19 compiles some representative anodic transmetalations:

Table 19. *Anodic transmetalations*

R_1ME	Anode (ME_2)	R_2ME	Ref.
Sodium tetraethyl-borate (H_2O)	Pb	$Pb(C_2H_5)_4$ 91%	613)
	Bi	$Bi(Et)_3$ 94%	613)
	Hg	$Hg(Et)_3$ 81%	613)
	Sb	$Sb(Et)_3$ 51%	613)
$NaF \cdot 2Al(Et)_3$	Pb	$Pb(Et)_4$ 100%	614)
$NaAl(Et)_3OC_4H_9$	Zn	Et_2Zn	615)
	Cd	Et_2Cd	615)
$NaAl(Me)_4$	Pb	$Pb(Me)_4$ 90%	616)

The electrolysis of methylmagnesium chloride in THF at a three dimensional anode consisting of lead pellets is the principal reaction of the NALCO-process, which produces 18.000 tons of $Pb(Me)_4$ per year [617]. Another technical process for the preparation of $Pb(Me)_4$ was developed by Ziegler and Lehmkuhl [618], electrolyzing $NaAl(Me)_4$ in diglyme at a lead anode.

Organometallics can be prepared by anodic generation of reactive metal ions. Alkylaluminium halides, $R-Al-X_2$, are produced by electrolysis of alkyl halides in the presence of AlX_3 at an aluminium anode. AlX is assumed to be the reactive intermediate, which inserts into the C—Hal bond to form product (Eq. (276)).

$$CH_3\text{-}I \xrightarrow{\text{Al-I}} CH_3\text{-Al-}I_2 \qquad 619)$$

$$CH_2Cl_2 \xrightarrow{\text{Al-Cl}} Cl_2\text{-Al-}CH_2\text{-Al-}Cl_2 \qquad 620) \qquad\qquad (276)$$

Formation of Me_3SnOAc from Me_4Sn is achieved in an indirect process by anodically generated Hg_2^{2+} according to Eq. (277)) [621].

$$2\ Me_4Sn\ +\ Hg_2(OAc)_2\ \longrightarrow\ 2\ Me_3SnOAc\ +\ Hg\ +\ Hg(Me)_2 \qquad (277)$$

$$\underset{-e}{\underbrace{}}$$

17.2. Cathodic Generation of Organometallics

Organometallics are formed at the cathode if transient radicals produced in reductions react with the active electrode. This occurs as a side reaction in cathodic coupling (Sect. 12.2, Eq. (185)) of carbonyl compounds, *e.g.*, of acetone [389], or of activated olefins, *e.g.*, of methyl vinyl ketone [413] or acrylonitrile. Furthermore, in cathodic cleavage (Sect. 13.2, Eq. (227)) of alkyl bromides or iodides organometallics are formed, *e.g.*, $ME(CH_2CH_2CN)_2$ (ME = Pb, Tl, Sn, Hg) [481], bis(*p*-substituted benzyl)mercury [485], or dicyclopropylmercury [489].

Lithium organic compounds were obtained by 2e-reduction of 1,1-diphenylethylene or tetraphenylethylene. The first olefin yielded quantitatively dilithio-1,1,4,4-tetraphenylbutane by dimerization of the intermediate radical anion (Eq. (244)) whereas the second formed dilithiotetraphenylethane [622].

Dessy *et al.* [629-632] have systematically studied the electrochemical redox behavior of more than 200 organometallic compounds with various ligands (phosphines, alkynes, CO, cyclopentadiene) covering the metals of all main groups and the transition metal series. Their reduction can be rationalized by the following general scheme (Eq. (278)):

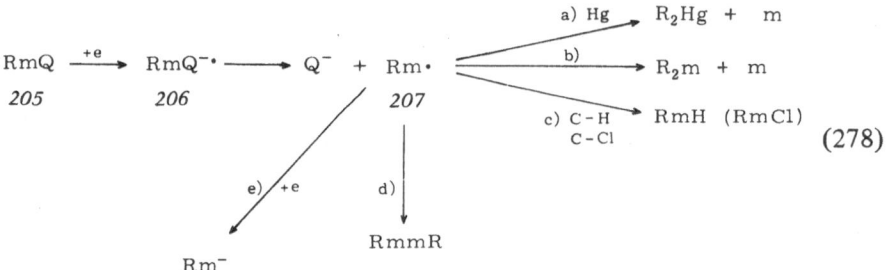

1e-reduction of the organometallic compound *205* yields a stable or transient radical anion [627]. The organometallic radical *207*, formed by dissociation of *206*, may transfer its ligands to mercury (path a), *e.g.*,

$$2(C_6H_5)_3Pb + 3\ Hg\ \rightarrow\ 3\ (C_6H_5)_2Hg + 2\ Pb\ ^{[623]}.$$

207 may furthermore disproportionate (path b) [633,634], abstract hydrogen (path c), *e.g.*, for *m* = Si, Ge, P [625,623], Cl [634], may dimerize (path d), *e.g.*, for *m* = Sn, As, Sb [623,625] or may be further reduced to Rm⁻ (path e). Rm⁻ can be also obtained by reductive cleavage of the Rm-mR bond [628] (Eq. (279)):

$$(Ph_3Sn(As, Sb, Bi))_2\ \xrightarrow{+e}\ 2\ Ph_3Sn(As, Sb, Bi)^- \qquad (279)$$

The use of these cathodically generated anions Rm⁻ for nucleophilic displacement of halogen in Rm'X offers an elegant synthesis for unsymmetric bimetallic species, *e.g.*,

$$Cp(CO)_2Fe\text{-}SnPh_3, \ Ph_3Pb\text{-}Fe(CO)_2Cp, \ Cp(CO)_3Mo\text{-}Sn\text{-}Ph_3 \ ^{630, \ 631)}$$

The relative nucleophilicities of Rm⁻ range in the order:

$$Ph_2Sb^-, Ph_2Bi^- = 10^{10}, Ph_3Sn^- = 10^8, C_6H_5S^- = 10^6, Re(CO)_5^- = 10^4, Co(CO)_4^- = 1 \ ^{629)}$$

18. Conclusion

The profits organic chemists can expect from electrochemistry in the future can best be expressed in the words of one of the world's leading experts in the field, J.O' M.Bockris writes in the article "Electrochemistry: The underdeveloped science" [641]: "electrochemistry nowadays ... has potentialities comparable with those faced by organic chemistry, say from 1920".

References

1) Basolo, F., Pearson, R.G.: Mechanisms of Inorganic Reactions, 2nd. ed. New York: John Wiley & Sons 1967

2) Reynolds, W.L., Lumry, R.W.: Mechanisms of Electron Transfer. New York: Ronald Press Company, 1966

3) Bilevich, K.A., Okhlobystin, O.Yu.: Russ. Chem. Rev. (English Transl.) *37*, 954 (1968)

4) Fichter, F.: Organische Elektrochemie. Dresden-Leipzig: Th. Steinkopf 1942

5) Brockmann, C.J.: Electro-organic Chemistry. New York: John Wiley and Sons 1926

6) Allen, M.J.: Organic Electrode Processes, Reinhold Publishing Corp., New York: 1958

7) Swann, S.: In: Technique of Organic Chemistry, Vol. II, 2nd ed. p. 385; A. Weissberger, Ed. New York: Interscience Publishers 1956

8) Meites, L.: In: Technique of Organic Chemistry, Vol. I, part IV, p. 3281; Weissberger, A., Ed. New York: Interscience Publishers 1960

9) Popp, F.D., Schultz, H.P.: Chem. Rev. *62*, 19 (1962)

10) Weinberg, N.L., Weinberg, H.R.: Chem. Rev. *68*, 449 (1968)

10a) Mann, Ch.K. and Barnes, K.K.: Electrochemical Reactions in Nonaqueous Systems. New York, Marcel Dekker, Inc., 1970

11) Sasaki, K., Newby, W.J.: J. Electroanal. Chem. *20*, 137 (1969)

12) Brown, O.R., Harrison, J.A.: J. Electroanal. Chem. *21*, 387 (1969)

13) Lund, H.: Österr. Chemiker-Ztg. *68*, 43 (1967); *68*, 152 (1967)

14) Eberson, L.: In: Chemistry of the Carboxyl Group, chap. 2; S., Patai, Ed. New York: Interscience Publishers 1969

15) Perrin, C.L.: Progr. Phys. Org. Chem. *3*, 220 (1965)

16) Tomilov, A.P.: Russ. Chem. Rev. (English Transl.) *30*, 639 (1961)

17) Fioshin, M.Ya.: Russ. Chem. Rev. (English Transl.) *32*, 30 (1963)

18) Utley, J.H.P.: Ann. Rep. Chem. Soc. (B) *65*, 231 (1968); *66*, 217

18a) Fleischmann, M., Pletcher, D.: Roy. Inst. Chem. Rev. *2*, 87 (1969)

19) Conway, B.E.: Theory and Principle of Electrode Processes. Ronald Press Company New York: 1965

20) Conway, B.E.: Rev. Pure Appl. Chem. *18*, 105 (1968)

21) Conway, B.E., Vijh, A.K.: Chem. Rev. *67*, 623 (1967)

22) Delahay, P., Tobias, C.W. (Eds.): Advan. Electrochem. and Electrochem. Eng. *1* (1961); *2* (1962); *3* (1963); *4* (1966); *5* (1967)

23) Bard, A.J. (Ed.): Electroanal. Chem. *1* (1966); *2* (1967); *3* (1969)

24) Zuman, P.: Substituent Effects in Organic Polarography. New York: Plenum Press 1967; cf. also Zuman, P: Fortschr. Chem. Forsch. *12*, 1 (1969); Progr. Phys. Org. Chem. *5*, 81 (1967)

25) Adams, R.N.: Electrochemistry at Solid Electrodes, New York: Marcel Dekker, Inc. 1969

25a) Vielstich, W.: Instrumentenk. *71*, 29 (1963)

26) Swann, S.: Trans. Electrochem. Soc. *69*, 385 (1936); *77*, 459 (1940); *88*, 103 (1947); Univ. Ill. Eng. Exptl. Sta. Bull. *34*, 17 (1936); *38*, 9 (1940); *44*, 42 (1947); Bibliography Electro-organic Chemistry Univ. Ill. Exptl. Sta. Bull. No. 50, *45*, 69; Electrochem.. Tech. *1*, 308 (1963); *5*, 53, 101, 393, 549 (1967)

27) Swann, S.: Electrochem. Tech. *4*, 550 (1966); *5*, 389, 467 (1967)

28) Baizer, M.M.: Naturwissenschaften *56*, 405 (1969)

29) Adams, R.N.: Accounts Chem. Res. *2*, 175 (1969)

30) Fleischmann, M.F., Pletcher, D.: Platinum Metals Rev. *13*, 46 (1969)

31) Ross, S.D.: Trans. N.Y. Acad. Sci., Ser. II *30*, 901 (1968)
32) Wawzonek, S.: Science *155*, 39 (1967)
33) Johnston, K.M.: Educ. Chem. *4*, 299 (1967); *5*, 15 (1968)
34) Hoijtink, G.J.: Chem. Ing.-Tech. *35*, 333 (1963)
35) Belleau, B., Au-Young, Y.K.: Can. J. Chem. *47*, 2117 (1969)
36) Schäfer, H., Steckhan, E.: Angew. Chem. (Intern. Ed. Engl.) *8*, 518 (1969)
37) Manning, G., Parker, V.D., Adams, R.N.: J. Am. Chem. Soc. *91*, 4584 (1969)
38) Parker, V.D., Eberson, L.: Tetrahedron Letters 2843 (1969)
39) Dickinson, T., Ovenden, P.J.: Chem. Brit. *5*, 260 (1969)
40) Adams, R.N.: J. Electroanal. Chem. *8*, 151 (1964);
40a) Balashova, N.A., Kazarinov, V.E.: Electroanal. Chem. *3*, 135 (1969)
40b) Kuwana, T., Strojek, J.W.: Discussions Faraday Soc. *45*, 134 (1968) and references cited therein. – T. Osa, Th. Kuwana, J. Electroanal. Chem. *22*, 389 (1969);
40c) Winograd, N., Kuwana, Th., J. Electroanal. Chem. *23*, 333, (1969); Hansen, W.N., Kuwana, Th., Osteryoung, R.: Anal. Chem. *38*, 1810 (1966)
40d) Janzen, E.G., Blackburn, B.J.: J. Am. Chem. Soc. *91*, 4481 (1969) and references cited therein;
40e) Fischer, H., Bargon, J.: Accounts Chem. Res. *2*, 110 (1969)
41) House, H.O.: Modern Synthetic Reactions, chap. 3. New York-Amsterdam: W.A. Benjamin, Inc. 1965
42) Andrulis, P.J. Jr., Dewar, M.J.S., Dietz, R., Hunt, R.L.: J. Am. Chem. Soc. *88*, 5473 (1966)
43) Heiba, E.I., Dessau, R.M., Koehl, W.J. Jr.: J. Am. Chem. Soc., *91*, 6830 (1969)
44) Fleischmann, M., Pletcher, D.: Tetrahedron Letters 6255 (1968);
44a) Suppliers of polarographs: Metrohm, Switzerland; Radiometer, Denmark; Heathkit, U.S.A.; see also advertisements in J. Electroanal. Chem.,Analytic. Chem., J. Electrochem. Soc.
45) Wilson, C.L., Lippincott, W.T.: J. Am. Chem. Soc. *78*, 4291 (1956); J. Electrochem. Soc. *103*, 672 (1956)
46) Linstead, R.P., Bunt, J.C., Weedon, B.C.L., Shephard, B.R.: J. Chem. Soc. 3624 (1952)
47) Harvey, D.R., Norman, R.O.C.: J. Chem. Soc. 4860 (1964)
48) Eberson, L., Nyberg, K.: J. Am. Chem. Soc. *88*, 1686 (1966); Acta Chem. Scand. *18*, 1568 (1964)
49) Leung, M., Herz, J., Salzberg, H.W.: J. Org. Chem. *30*, 310 (1965). – Salzberg, H.W., Leung, M.: J. Org. Chem. *30*, 2873 (1965)
50) Ross, S.D., Finkelstein, M., Petersen, R.C.: J. Am. Chem. Soc. *86*, 4139 (1964)
51) Eberson, L., Olofsson, B.: Acta Chem. Scand. *23*, 2355 (1969); for a conflicting view, see Parker, V.D., Burgert, B.E.: Tetrahedron Letters 2411 (1968)
52) Parker, V.D., Eberson, L.: Chem. Commun. 340 (1969)
53) Parker, V.D., Chem. Commun. 848 (1969)
54) Baizer, M.M.: J. Org. Chem. *29*, 1670 (1964)
55) Koyama, K., Susuki, T., Tsutsumi, S.: Tetrahedron Letters 627 (1965)
56) Parker, V.D., Burgert, B.E.: Tetrahedron Letters 4065 (1965)
57) Eberson, L., Nilsson, S.: Discussions Faraday Soc. *45*, 242 (1968)
58) Tsutsumi, S., Koyama, K.: Trans. Faraday Soc. *45*, 247 (1968)
59) Susuki, T., Koyama, K., Omori, A., Tsutsumi, S.: Bull. Chem. Soc. Japan *41*, 2663 (1968)
60) Andreades, S., Zahnow, E.W.: J. Am. Chem. Soc. *91*, 4181 (1969)
61) Parker, V.D., Burgert, B.E.: Tetrahedron Letters 3341 (1968)
62) Weinberg, N.L., Brown, E.A.: J. Org. Chem. *31*, 4058 (1966)
63) – J. Org. Chem. *33*, 4326 (1968); cf. also Smith, P.J., Mann, C.K.: J. Org. Chem.*33*, 316 (1968)
64) Meites, L.: Pure Appl. Chem. *18*, 35 (1969)
65) Peover, M.E.: Electroanal. Chem. *2*, 1 (1967); cf. also Ref. 34)
65a) Michielli, R.F., Elving, P.J.: J. Am. Chem. Soc. *90*, 1989 (1968);
65b) Sadler, J.L., Bard, A.J.: J. Am. Chem. Soc. *90*, 1970 (1968)

66) For discussion and references, see Zahradnik, R., Parkanyi, C.: Talanta *12*, 1289 (1965)

66a) For an extensive review on nonaqueous solvents in electrochemistry, see Mann, C.K.: Electroanal. Chem. *3*, 57 (1969) Bard, A.J. (Ed.)

67) Billon, J.P.: Bull. Soc. Chim. France 863 (1962)

68) Courtot-Coupez, J., Le Demezet, M.: Bull. Soc. Chim. France 4744 (1967)

69) Dubois, J.E., Lacaze, P.C., de Ficquelmont, A.M.: Compt. Rend. *262* C, 181 (1966)

70) Quoted from Ref. [25], - p. 33

71) Cauquis, G., Serve, D.: Bull. Soc. Chim. France 302 (1966)

72) Nelson, R.F., Adams, R.N.: J. Electroanal. Chem. *13*, 184 (1967)

73) Coetzee, J.F., Simon, J.M., Bertozzi, R.J.: Anal. Chem. *41*, 766 (1969)

74) Perichon, J., Buvet, R.: Bull. Soc. Chim. France 1279 (1968)

75) Swain, C.G., Ohno, A., Roe, D.K., Brown, R., Maugh, T.: J. Am. Chem. Soc. *89*, 2648 (1967)

75a) Woodhall, B.J., Davies, G.R.: Abstracts, Meeting of the Society for Electrochemistry, England, April 15-16, 1969, p. 26

76) Baizer, M.M.: J. Electrochem. Soc. *111*, 215 (1964)

77) Ercoli, R., Guainazzi, M., Silvestri, G.: Chem. Commun. 927 (1967)

78) Nyberg, K.: Acta Chem. Scand., *24*, 1609 (1970)

79) For a discussion and an extensive list of references, see: Parsons, R.: Rev. Pure Appl. Chem. *18*, 91 (1968)

80) Beck, F.: Ber. Bunsenges. Physik. Chem. *72*, 379 (1968)

81) Nyberg, K.: Chem. Commun. 774 (1969); cf. also Ref. [51]

82) Koehl, W.J., Jr: J. Am. Chem. Soc. *86*, 4686 (1964). For a case where substitution of carbon for platinum does not change the product composition, see Eberson, L., Nilsson, S.: Acta Chem. Scand. *22*, 2453 (1968)

83) Ross, S.D., Finkelstein, M.: J. Org. Chem., *34*, 2923 (1969)

84) Hartley, A.M., Axelrod, H.D.: J. Electroanal. Chem. *18*, 115 (1968)

85) Woolford, R.G.: Can. J. Chem. *40*, 1846 (1962); cf. also Woolford, R.G., Arbis, W., Rosser, A.: Can. J. Chem. *42*, 1788 (1964)

86) Barry, C., Cauquis, G.: Bull. Soc. Chim. France 1032 (1966)

87) Peltier, D., Le Guyader, M., Tacussel, J.: Bull. Soc. Chim. France 2609 (1963). Moinet, C., Peltier, D.: Bull. Soc. Chim. France 690 (1969);

87a) Iversen, P.E., Lund, H.: Acta Chem. Scand. 19, 2303 (1965)

88) Wright, C.M., Levering, D.R.: Tetrahedron *19*, 3 (1963)

89) Sioda, R.E.: Electrochim. Acta *13*, 375 (1968) and references cited therein;

89a) Suppliers of ion exchange membranes: American Machine & Foundary Co., Stamford, Conn., U.S.A.; Ionics Chemical Co., Cambridge, Mass., U.S.A., Asahi Chemical Industry Co., Tokyo, Japan;

89b) Suppliers of potentiostats: Chemical Electronics, Durham, England; Amel, Milan, Italy; Tage Juul Electronics, Herlev, Denmark; Elron, Tel-Aviv, Israel

90) Bewick, A., Brown, O.R.: J. Electroanal. Chem. *15*, 129 (1967)

91) Fleischmann, M., Goodridge, F.: Discussions Faraday Soc. *45*, 254 (1968) and references cited therein.

92) Hickling, A., Wilkins, R.: Discussions Faraday Soc. *45*, 261 (1968)

93) Allen, M.J.: Electrochem. Tech. *1*, 315 (1963)

94) MacMullin, R.B.: Electrochem. Tech. *2*, 106 (1964)

95) Goodridge, F.: Chem. Proc. Eng., February, 93 (1968), March 121 (1966)

96) Danly, D.E.: Hydrocarbon Processing, June, 159 (1966)

97) For a conflicting view on the merits of fluidized-bed electrodes, see Armstrong, R.D., Brown, O.R., Giles, R.D., Harrison, J.A.: Nature *219*, 94 (1968)

98) Witts, W.S.: Abstracts, Meeting of the Society for Electrochemistry (England), Thornton, April 15-16, 1968;

98a) Backhurst, J.R., Coulson, J.M., Goodridge, F., Plimley, R.E., Fleischmann, M.: J. Electrochem. Soc. *116*, 1600 (1969)

99) Elving, P.J., Pullman, B.: Advan. Chem. Phys. *3*, 1 (1961)

100) Mango, F.D., Bonner, W.A.: J. Org. Chem. *29*, 1367 (1964)

References

101) Koyama, K., Ebara, T., Tani, T., Tsutsumi, S.: Can. J. Chem. *47*, 2484 (1969)
102) Eberson, L., Nyberg, K., Finkelstein, M., Petersen, R.C., Ross, S.D.; Uebel, J.J.: J. Org. Chem. *32*, 16 (1967)
103) Paquette, L.A., Malpass, J.R., Barton, T.J.: J. Am. Chem. Soc. *91*, 4714 (1969) and references cited therein
104) Inoue, T., Koyama, K., Matsuoka, T., Tsutsumi, S.: Bull. Chem. Soc. Japan *40*, 162 (1967)
105) Ross, S.D., Finkelstein, M., Uebel, J.J.: J. Org. Chem. *34*, 1018 (1969)
106) Bernhardsson, E., Eberson, L., Sternerup, H.: Extended Abstracts, Meeting of the Electrochemical Society, New York, May 4-9, 1969, p. 311; configurational assignments of the two diastereomers have been revised
107) Horner, L., Röder, H.: Liebigs Ann. Chem. *723*, 11 (1969)
108) Dietz, R., Peover, M.E.: Discussions Faraday Soc. *45*, 154 (1968)
108a) Fry, A.J., Moore, R.H.: J. Org. Chem. *33*, 1283 (1968);
108b) Erickson, R.E., Annino, R., Scanlon, M.D., Zon, G.: J. Am. Chem. Soc. *91*, 1767 (1969);
108c) See also Conway, B.E., Rudd, E.J., Gordon, L.G.M.: Extended Abstracts, Meeting of the Electrochemical Society, New York, May 4-9, 1969, p. 340;
108d) Gouwley, R.N., Grimshaw, J., Miller, P.G.: Chem. Commun. *3*, 1278 (1967);
108e) Horner, L., Degner, D.: Tetrahedron Letters 1968, 5889
109) Parker, V.D., Nyberg, K., Eberson, L.: J. Electroanal. Chem. *22*, 150 (1969)
110) Fritsch, J.M., Weingarten, H.: Extended Abstracts, Meeting of the Electrochemical Society, New York, May 4-9, 1969, p. 346
111) Bard, A.J., Phelps, J.: J. Electroanal. Chem. *25*, App. 2-5 (1970)
111a) For a review on the chemistry of radical anions, see Swarc, M.: Progr. Phys. Org. Chem. *6*, 323 (1968). – Dorfmann, L.M.: Accounts Chem. Res. *3*, 224 (1970)
112) Staples, T.L., Jagur-Grodzinski, J., Swarc, M.: J. Am. Chem. Soc. *91*, 3721 (1969)
113) Hoijtink, G.J.: Ind. Chim. Belge *12*, 1371 (1963)
114) Velthorst, N.H., Hoijtink, G.J.: J. Am. Chem. Soc. *87*, 4529 (1965)
115) – – J. Am. Chem. Soc. *89*, 209 (1967)
116) Kastening, B.: Electrochim. Acta *9*, 241 (1964). – Kastening, B., Vavricka, S.: Ber. Bunsenges. Physik. Chem. *72*, 27 (1968)
117) Kolb, D., Wirths, W., Gerischer, H.: Ber. Bunsenges. Physik. Chem. *73*, 148 (1969)
118) Sioda, R.E.: J. Phys. Chem. *72*, 2322 (1968)
119) Shine, H.J., Murata, Y.: J. Am. Soc. *91*, 1872 (1969); J. Org. Chem. *34*, 3368 (1969)
120) Lund, H.: Acta Chem. Scand. *11*, 491 (1957)
121) Fichter, F., Ris, H.: Helv. Chim. Acta *7*, 803 (1924)
122) Brit. Patent 1, 051, 614; Chem. Abstr. *66*, 81920 (1967)
123) See Ref. 10), Table XVII
124) U.S. Patent 3,200,054; Chem. Abstr. *63*, 11023 (1965)
125) Fichter, F., Sjostedt, P.: Chem. Ber. *43*, 3422 (1910)
126) U.S. Patent 2,521,147; Chem. Abstr. *44*, 10556 (1950)
127) German Patent 1,034,171; Chem. Abstr. *54*, 11775 (1960); cf. also Fichter, F., Wenk, W.: Chem. Ber. *45*, 1373 (1912)
128) Fichter, F., Lotter, P.: Helv. Chim. Acta *8*, 438 (1925)
129) Cauquis, G., Genies, M.: Tetrahedron Letters 3537 (1968)
130) Zuman, P., Barnes, D., Ryvolova-Kejharova, A.: Discussions Faraday Soc. *45*, 202 (1968) and references cited therein
131) Wawzonek, S., Gundersen, A.: J. Electrochem. Soc. *107*, 537 (1960)
132) Evans, D.H., Woodbury, E.C.: J. Org. Chem. *32*, 2158 (1967)
133) Stocker, J.H., Jenevein, R.M.: Symposium Abstract, "The Synthetic and Mechanistic Aspects of Electroorganic Chemistry", U.S. Army Research Office, Durham, N.C., p. 221
134) Kabasakalian, P., McGlotten, J., Basch, A., Yudis, M.D.: J. Org. Chem. *26*, 1738 (1961)
135) Mandell, L., Powers, R.M., Day, R.A.: J. Am. Soc. *80*, 5284 (1958)
136) Throop, L., Tökes, L.: J. Am. Soc. *89*, 4789 (1967)

137) Lund, H.: Acta Chem. Scand. *17*, 972 (1963)
138) Iversen, P.E., Lund, H.: Acta Chem. Scand. *21*, 279 (1965)
139) Acta Chem. Scand. *21*, 389 (1967)
140) Mettler, C.: Chem. Ber. *38*, 1745 (1905); *39*, 2933 (1906);
140a) Carrahar, P., Drakesmith, F.G.: Chem. Commun. 1562 (1968)
141) Lund, H.: Acta Chem. Scand. *17*, 2325 (1963)
142) – Acta Chem. Scand. *13*, 249 (1959)
143) – Acta Chem. Scand. *18*, 563 (1964)
144) – Tetrahedron Letters 3651 (1968)
144a) Fry, A.J., Newberg, J.H.: J. Am. Chem. Soc. *89*, 6374 (1967); – Reed, R.G.: J. Am. Chem. Soc. *91*, 6448 (1969)
145) Lund, H., Discussions Faraday Soc. *45*, 193 (1968)
146) Volke, J., Kardos, A.M.: Collection Czech. Chem. Commun. *33*, 2560 (1968)
147) Manousek, O., Zuman, P.: Chem. Commun. 158 (1965)
148) Arapakos, P.G., Scott, M.K.: Tetrahedron Letters 1975 (1968)
149) Iversen, P., Lund, H.: Tetrahedron Letters 4027 (1967)
150) Sayo, H., Tsukitani, Y., Masui, M.: Tetrahedron *24*, 1717 (1968)
151) Sayo, H., Masui, M.: Tetrahedron *24*, 5075 (1968)
152) Tallec, A.: Compt. Rend. *262 C*, 1886 (1966); *263 C*, 722 (1966); Ann. Chim. (Paris) *3*, 164 (1968)
153) Cadle, S.H., Tice, P.R., Chambers, J.Q.: J. Phys. Chem. *71*, 3517 (1967)
154) Hranilovic, J., Koruncev, D., Gustak, E.: Electrochem. Tech. *6*, 62 (1968)
155a) Lund, H.: Extended Abstracts, Symposium on the Synthetic and Mechanistic Aspects of Electroorganic Chemistry, U.S. Army Research Office, Durham, N.C., U.S.A., October 14-16, 1968, p. 197;
155b) – In: The chemistry of the carbon-nitrogen double bond, p. 505; Ed. S. Patai, Interscience Publishers 1970
156) Oelschläger, H., Hoffmann, H.: Arch. Pharm. *300*, 817 (1967)
157) Ramaswamy, R., Venkatachalapathy, M.S., Udupa, H.V.K.: J. Electrochem. Soc. *110*, 202 (1963)
158) Rausch, M.D., Popp, F.D., McEwen, W.E., Kleinberg, J.: J. Am. Chem. Soc. *76*, 3522 (1954)
159) – J. Org. Chem. *25*, 1186 (1960)
160) Staudinger, H., Ullrich, V.: Z. Naturforsch. *19b*, 409 (1964). – Ullrich, V., Hey, D., Zubrzycki, Z., Staudinger, H.: Z. Naturforsch. *20b*, 1185 (1965)
161) Wright, C.M., Heck, R.F., Yokoyama, W.: Extended Abstracts, Meeting of the Electrochemical Society, New York, May 4-9, 1969, p. 361
162) Arad, Y., Levy, M., Miller, I.R., Vofsi, D.: J. Electrochem. Soc. *114*, 899 (1967)
163) Rosen, H., Arad, Y., Levy, M., Vofsi, D.: J. Am. Chem. Soc. *91*, 1425 (1969)
164) Littlehailes, J.D., Woodhall, B.J.: Discussions Faraday Soc. *45*, 187 (1968)
165) Dietz, R., Peover, M.E., Rothbaum, H.P.: Abstracts, Symposium on the Synthetic and Mechanistic Aspects of Electroorganic Chemistry, U.S. Research Army Office, Durham, N.C., U.S.A., October 14-16, 1968, p. 137
166) – Abstracts, Meeting of the Society for Electrochemistry, England, April 15-16, 1969, p. 8
166a) Dietz. R., Peover, M.E., Rothbaum, P.: Chem.-Ing.-Tech. *42*, 185 (1970)
167) Iversen, P.E., Lund, H.: Tetrahedron Letters 3523 (1969)
167a) Shono, T., Mitani, M.: J. Am. Chem. Soc. *90*, 2728 (1968)
168) Nagase, S.: Fluorine Chem. Rev. *1*, 77 (1967)
169) Burdon, J., Tatlow, J.C.: Advan. Fluorine Chem. *1*, 129 (1960)
170) Cornforth, J.W., Jones, F., Clifford, K.H., Green, D.T.: Abstracts, Meeting of the Society for Electrochemistry, England, April 15-16, 1969, p. 6
171) Ibl, N., Selvig, A.: Abstracts, Meeting of Applied Electrochemistry, Bonn, Germany, October 6-8, 1969; Chem.-Ing.-Tech. *42*, 180 (1970)
172) Wawzonek, S., Duty, R.C.: J. Electrochem. Soc. *108*, 1135 (1961)
173) Wawzonek, S., Wagenknecht, J.H.: J. Electrochem. Soc. *110*, 420 (1963)

References

174) Su, T.: Dissertation, The University of Iowa, U.S.A., 1967; Univ. Microfilms No. 67-9111
175) Schmeisser, M., Sartori, P.: Chem.-Ing.-Tech. *36*, 9 (1964)
176) Forche, E., Methoden der Organischen Chemie (Houben-Weyl), Stuttgart: G.Thieme Verlag 1962. Ed.E. Müller, Vol. V,3, p. 38
177) Voitovich, Ya.N., Kazakov, V.Ya., Kozyreva, A.N., Levin, A.: Zh.Prikl. Khim. *42*, 131 (1969); Chem. Abstr. *70*, 120545 (1969)
178) Simons, J.H., and coworkers: Trans. Elechtrochem.Soc. *95*, 47 (1949)
179) Nagase, S., Baba, H., Abe, T.: Bull.Chem.Soc. Japan *39*, 2304 (1966)
180) Goerig, D., Jonas, H., Moschel, W.: DBP 1040009 (1952), C. 1956, 10598
181) Nagase, S., Abe, T., Babe, H.: Chem. Abstr. *70*, 33773 (1969)
182) Scholberg, H.M., Brice, H.G.: AP 2717871, Chem. Abstr. *49*, 15572h (1955)
183) Nagase, S., Abe, T., Babe, H.: Bull. Chem. Soc. Japan *41*, 1921 (1968)
184) Ryabinin, N.A., Kolenko, I.P., Lunadin, B.N., Butina, I.V.: Chem. Abstr. *70*, 68024 (1969)
185) Kolenko, I.P., Ryabinin, N.A., Lundin, B.N.: Chem.Abstr. *70*, 87422 (1969)
186) Kauck, E., Simons, J.H.: Chem. Abstr. *46*, 6015a (1952); AP 2, 594, 272
187) Gramstadt, T., Haszeldine, R.N.: J. Chem. Soc. *1956*, 173; ibid. *1957*, 2640
188) Dow Corning Corp.: Brit. Pat. 1.099.240; Chem. Abstr. *68*, 83821w (1968)
189) Simmons, T.C., Hoffmann, F.W.: J.Am. Chem. Soc. *79*, 3429 (1957)
190) Kauck, E.A., Simons, J.H.: C. 1955, 4937
191) Shono, T., Matsamura, Y.: J.Am. Chem. Soc. *91*, 2803 (1969)
192) – Kosaka, T.: Tetrahedron Letters *1968*, 6207
193) Ross, S.D., Finkelstein, M., Petersen, R.C.: J. Am. Chem. Soc. *88*, 4657 (1966)
194) Shono, T., Matsumara, Y.: J. Am. Chem. Soc. *90*, 5937 (1968)
195) Cooper, A., Mantell, Ch.L.: Chem. Abstr. *65*, 3323f (1966)
196) Kharitinov, N.P., Nechaev, B.P., Fedorova, G.T.: Zh. Obshch. Khim. *39*, 824 (1969); Chem. Abstr. *71*, 56040w (1969)
197) Takeda, A., Wada, S., Torii, S., Matsui, J.: Bull. Chem. Soc. Japan *42*, 1047 (1969). - Takeda, A., Torii, S., Oka, N.: Tetrahedron Letters *1968*, 1781
198) Eberson, L.: Acta Chem. Scand. *17*, 2004 (1963)
199) Woolford, R.G., Song, J., Lin, W.S.: Can. J. Chem. *45*, 1837 (1967)
200) Traynham, J.G., Dehn, J.S.: J. Am. Chem. Soc. *89*, 2139 (1967)
201) Skell, P.S., Starer, I.: J.Am. Chem. Soc. *81*, 4117 (1959); ibid. *82*, 2971 (1960); ibid. *84*, 3962 (1962)
202) Ref. [14), Table 9 and 7
203) Corey, E.J., Bauld, N.L., La Londe, R.T., Casanova, J. Jr., Kaiser, E.T.: J. Am. Chem. Soc. *82*, 2645 (1960)
204) Reichenbacher, P.H., Morris, M.D., Skell, P.S.: J. Am. Chem. Soc. *90*, 3432 (1968)
205) Gassman, P.G., Zalar, F.V.: J. Am. Chem. Soc. *88*, 2252 (1966)
206) Shono, T., Nishiguchi, J., Yamane, S., Oda, R.: Tetrahedron Letters 1969, 1965
207) Skell, P.S., Reichenbacher, P.H.: J. Am. Chem. Soc. *90*, 2309 (1968)
208) – J. Am. Chem. Soc. *90*, 3436 (1968)
209) Wharton, P.S., Hiegel, G.A., Coombs, R.V.: J. Org. Chem. *28*, 3217 (1963)
210) Corey, E.J., Sauers, R.R.: J. Am. Chem. Soc. *81*, 1743 (1959)
211) Binns, T.D., Brettle, R., Cox, G.B.: J. Chem. Soc. C *1969*, 2499
212) Garwood, R.F., Din, N.U., Weedon, B.C.L.: Chem. Commun. *1968*, 923
213) Bunyan, P.J., Hey, D.H.: J. Chem. Soc. *1962*, 324, 2771
214) Smets, G., Borght, X. van der, Hawen, G.J. van: J. Polymer Sci. A*2*, 5187 (1964)
215) Finkelstein, M, Petersen, R.C.: J. Org. Chem. *25*, 136 (1960)
216) Wladislaw, B., Ayres, A.M.J.: J. Org. Chem. *27*, 281 (1962)
217) Linstead, R.P., Shephard, B.R., Weedon, B.C.L.: J. Chem. Soc. *1951*, 2854
218) Gassman, P.G., Fox, B.L.: J. Org. Chem. *32*, 480 (1967)
219) Fichter, F., Herndl, J.: Helv. Chim. Acta 25, 229 (1942)
220) –, W. Steinbuch: Helv. Chim. Acta 26, 695 (1943)
221) –, Bloch, E.: Helv. Chim. Acta 22, 1529 (1939)

222) Eberson, L., Nyberg, K.: Acta Chem. Scand. *18*, 1567 (1964)
223) Eberson, L.,: J. Org. Chem. *27*, 2329 (1962)
224) Muck, D.L., Wilson, E.R.: Extended abstracts, Meeting of the Electrochemical Society, New York, May 4-9, 1969, Abstr. No. 129, p. 308
225) Kornprobst, J.M., Laurent, A., Laurent-Dieuziede, E.: Bull. Soc. Chim. France 1968, 3657
226) Eberson, L.: J. Am. Chem. Soc. *89*, 4669 (1967)
227) Parker, V.D., Burgert, B.: Tetrahedron Letters *1965*, 4065
228) Inoue, T., Koyama, K., Tsutsumi, S.: Bull. Chem. Soc. Japan *37*, 1597 (1964); ibid. *38*, 661 (1965); Tetrahedron Letters *1963*, 1409
229) Koyama, K., Suzuki, T., Tsutsumi, S.: Tetrahedron Letters *1965*, 627
230) Sasaki, K., Urata, H., Umeyama, K., Nagaura, S.: Electrochim. Acta *12*, 137 (1967)
231) Millington, J.P.: J. Chem. Soc. B *1969*, 982;
231a) Parker, V.D., Eberson, L.: Tetrahedron Letters *1969*, 2839
232) Ref. $^{10)}$, Table 3
232a) Koyama, K., Yoshida, K., Tsutsumi, S.: Bull. Chem. Soc. Japan *39*, 516 (1966)
233) Tsutsumi, S., Koyama, K., Ebara, T.: Bull. Chem. Soc. Japan *41*, 2668 (1968)
234) Koehl, W.J. Jr.: J. Org. Chem. *32*, 614 (1967)
235) Susuki, T., Koyama, K., Omari, A., Tsutsumi, S.: Bull. Chem. Soc. Japan *41*, 2663 (1968)
236) Yoshida, K., Fueno, T.: Chem. Commun. *1970*, 711
237) Fichter, F., Schönmann, P.: Helv. Chim. Acta *19*, 1411 (1936)
238) Sasaki, K., Takehira, U., Shiba, H.: Electrochim. Acta *13*, 1623 (1968)
239) Covitz, F.H.: Chem. Abstr. *71*, 91080 (1969); Fr. Pat. 1,544,350
240) Miller, L.L., Kujawa, E.P., Campbell, C.B.: J. Am. Chem. Soc. *92*, 2821 (1970)
241) Parker, V.D., Adams, R.N.: Tetrahedron Letters *1969*, 1721
242) Eberson, L., Olofson, B.: Acta Chem. Scand. *23*, 2355 (1969)
243) —, Nyberg, K.: Tetrahedron Letters *1966*, 2389
244) Ross, S.D., Finkelstein, M., Petersen, R.C.: J. Am. Chem. Soc. *89*, 4088 (1967)
245) Tsutsumi, S., Koyama, K., Susuki, T.: Tetrahedron *23*, 2665 (1967)
246) Wladeslaw, B., Viertler, H.: Chem. Ind. (London) *1965*, 39
247) Weinberg, N.L.: Extended Abstracts, Symposium on the Synthetic and Mechanistic Aspects of Electroorganic Chemistry, U.S., Army Research Office, Durham, N.C. U.S.A. October 14-16, 1968, p. 263
248) —, Reddy, T.B.: J. Am. Chem. Soc. *90*, 91 (1968)
249) Basselier, J.J., Cauquis, G., Cros, J.L.: Chem. Commun. *1969*, 1171
250) Hudson, A.G., Pedler, A.E., Tatlow, J.C.: Tetrahedron Letters *1968*, 2143
251) Wawzonek, S., Gundersen, A.: J. Electrochem. Soc. *107*, 537 (1960)
252) Wawzonek, S., Berkey, R., Blake, E.W., Runner, M.E.: J. Electrochem.Soc. *103*, 456 (1956)
253) McDowell, C.S.: Diss. Abstr. B, 28 (6) 2348 (1967)
254) Rifi, M.R.: J.Am. Chem. Soc. *89*, 4442 (1967)
255) —, Extended abstracts, Meeting of the Electrochemical Society, New York, May 4-9, 1969, p. 320
256) Parker, V.D.: J. Am. Chem. Soc. *91*, 5380 (1969)
257) Lindsey, R.V., Peterson, M.L.: J. Am. Chem. Soc. *81*, 2073 (1959)
258) Smith, W.B., Gilde, H.: J. Am. Chem. Soc. *81*, 5325 (1959)
259) Smith, W.B., Gilde, H.: J. Am. Chem. Soc. *83*, 1355 (1961)
260) Khrizolitova, M.A., Mirkind, L.A., Fioshin, M.Y.: Zh. Org. Khim. *6*, 219 (1970); Dokl. Akad. Nauk SSSR *1968* (182), 617; *69*, 112842 g (1968); *71*, 26969 p (1969); Zh. Org. Khim. *1968*, 1705; Chem. Abstr. *70*, 19497, (1969)
261) Smith, W.B., Gilde, H.: J. Am. Chem. Soc. *82*, 659 (1960)
262) Goldschmidt, S., Stöckl, E.: Chem. Ber. *85*, 630 (1952)
263) Fioshin, M.Y., Kamneva, A.I., Mirkind, L.A., Salmin, L.A., Kornienko, A.G.: Chem. Abstr. *58*, 11205 (1963)
264) Schäfer, H., Alazrak, A.: Angew. Chem. *80*, 485 (1968); Angew. Chem. Intern. Ed. Engl. *7*, 474 (1968)

References

264a) Whalling, Ch., Huyser, E.S.: In: A.C. Cope(ed.), Organic Reactions, Vol. 13 pp.
91-149. New York: J. Wiley & Sons, Inc. 1963
265) Schäfer, H.: Chem.-Ing.-Tech. *41*, 179 (1969)
265a) Lawesson, S.O., Sosnovsky, G.: Svensk Kem. Tidskr. *75*, 343 (1963); *75*, 568 (1963)
266) Schäfer, H.: Angew. Chem. *82*, 134 (1970); Angew. Chem. Intern. Ed. Engl. *9*, 158 (1970)
267) –, Küntzel, H.: Tetrahedron Letters, 3333 (1970)
268) –, Chem.-Ing.- Tech. *42*, 164 (1970
269) –, Steckhan, E.: unpublished
270) Foita, G., Fleischmann, M., Pletcher, D.: J. Electroanal. Chem. *25*, 455 (1970)
270a) Weinberg, N.L., Hoffmann, A.K.: 21. st CITCE Meeting, Prag 1970, Extended abstracts p. 467
271) Leininger, R., Pasiut, L.A.: Trans. Electrochem. Soc. *88*, 73 (1945)
272) Clauson-Kaas, N., Limborg, F., Glens, K.: Acta Chem. Scand. *6*, 531 (1952)
273) – –, Dietrich, P.: Acta Chem. Scand. *6*, 545 (1952)
274) Murakami, M., Chen, J.: Bull. Chem. Soc. Japan *36*, 263 (1963)
275) Ponomarev, A.A., Markushina, I.A.: J. Gen. Chem. USSR *33*, 3892 (1963)
276) Yoshida, K., Fueno, T.: Bull. Chem. Soc. Japan *42*, 2411 (1969)
277) Belleau, B., Weinberg, N.L.: J. Am. Chem. Soc. *85*, 2525 (1963)
278) Quoted from Ref.
279) Weinberg, N.L., Brown, E.A.: J. Org. Chem. *31*, 4054 (1966)
280) Vermillion, F.J. Jr., Pearl, I.A.: J. Electrochem. Soc. *111*, 1392 (1964)
281) Dimroth, K., Perst, H., Schlömer, K., Worschech, K., Müller, K.H.: Chem. Ber. *100*, 629 (1967)
282) Scott, A.J., Dodson, P.A., Mc.Capra, F., Meyers, M.B.: J. Am. Chem. Soc. *85*, 3702 (1963). – Iwasaki, H., Cochen, L.A.,Witkop, B.: J. Am. Chem. Soc. *85*, 3701 (1963)
283) Lund, H.: Acta Chem. Scand. *11*, 1323 (1957)
284) Parker, V.D., Eberson, L.: Chem. Commun. *1969*, 451
285) Brettle, R., Baggaly, A.J.: J. Chem. Soc. C, *1968*, 2055
286) Courbis, P., Guillemont, A.: Compt. Rend. Ser. C *262*, 1435 (1966)
287) Inoue, T., Tsutsumi, S.: Bull. Chem. Soc. Japan *38*, 661 (1965)
288) –, Koyama, K., Matsuoka, T., Matsuoka, K., Tsutsumi, S.: Tetrahedron Letters *1963*, 1409
289) Inoue, T., Tsutsumi, S.: J. Am. Chem. Soc. *87*, 3525 (1965)
290) Dye, J.L.: Accounts Chem. Res. *1*, 306 (1968); Adv. Chem. Series of A.C.S. *50* (1965)
291) Osa, T., Yamagishi, T., Kodama, T., Misono, A.: Extended abstracts, Symposium on the Synthetic and Mechanistic Aspects of Electroorganic Chemistry, U.S. Army Research Office, Durham, N.C., U.S.A., October 14-16, pp. 157; Bull. Chem. Soc. Japan *41*, 69 (1968); J. Electrochem. Soc. *115*, 266 (1968)
292) Misono, A., Osa, T., Yamagishi, T.: Bull. Chem. Soc. Japan *40*, 427 (1967)
293) Hart, E.J., Gordon, S., Fielden, E.M.: J. Phys. Chem. *70*, 150 (1966)
294) Sternberg, H.W., Markby, R.E., Wender, I., Mohilner, D.M.: J. Am. Chem. Soc. *89*, 186 (1967); Extended abstracts, Symposium on the Synthetic and mechanistic Aspects of Electroorganic Chemistry, U.S. Army Research Office, Durham, N.C., U.S.A., October 14-16, 1968, pp. 179. – Sternberg, H.W., Mohilner, D.M.: J. Am. Chem. Soc. *91*, 4191 (1969)
294a) Caterall, R., Stodulski, L.P., Symons, M.C.R.: J. Chem. Soc. A *1968*, 437
295) Asahara, T., Seno, M., Kaneko, H.: Bull. Chem. Soc. Japan *41*, 2985 (1968)
296) Hoijtink, G.J.: Rec. Trav. Chim. *76*, 885 (1957)
297) Benkeser, R.A., Kaiser, E.M.: J. Am. Chem. Soc. *85*, 2858 (1963). – Benkeser, R.A., Kaiser, E.M., Lambert, R.F.: J. Am. Chem. Soc. *86*, 5272 (1964)
298) –, Tincher, C.A.: J. Org. Chem. *33*, 2727 (1968)
299) –, Watanabe, H., Mels, S.J., Sabol, M.: J. Org. Chem. *35*, 1210 (1970)
300) Throop, L.J.: U.S. Patent 3,444,057, C.A. *71*, 50359 k (1969)
301) Horner, L., Röder, H.: Liebigs Ann. Chem. *723*, 11 (1969)
302) Rosenthal, I., Hayes, J.R., Martin, A.J., Elving, Ph.J.: J. Am. Chem. Soc. *80*, 3050 (1958)
303) Lee, J.B., Cashmore, P.: Chem. Ind. London *1966*, 1758

304) Wawzonek, S., Gundersen, A.: J. Electrochem. Soc. *107*, 537 (1960)
305) —, Duty, R.C., Wagenknecht, J.H.: J. Electrochem. Soc. *111*, 74 (1964)
306) —, Gundersen, A.: J. Electrochem. Soc. *111*, 324 (1964)
307) —, Blaha, E.W., Berkey, R., Runner, M.E.: J. Electrochem. Soc. *102*, 235 (1955)
308) Takeda, K., Igarashi, K., Narisoda, M.: Steroids *4*, 305 (1964)
309) Plieninger, H., Lehnert, W.: Chem. Ber. *100*, 2427 (1967)
310) Westberg, H.H., Dauben, H.J., Jr.: Tetrahedron Letters *1968*, 5123
311) Radlick, Ph., Klem, R., Spurlock, S., Simms, J.J., Tamelen, E.E. van, Whitesides, T.: Tetrahedron Letters *1968*, 5117
312) Märkl, J., Schubert, H.: Tetrahedron Letters *1970*, 1273
313) Fichter, F., Petrovitch, A.: Helv. Chim. Acta *24*, 549 (1941)
314) Corey, E.J., Casanova, J., Jr.: J. Am. Chem. Soc. *85*, 165 (1963)
315) Rosenthal, J., Lacoste, R.J.: J. Am. Chem. Soc. *81*, 3268 (1959)
316) Feoktistov, L.G., Tomilov, A.P., Gol'din, M.M.: Izv. Akad. Nauk. SSSR, Ser. Khim. *1963*, 1352; Chem. Abstr. *59*, 12624 f (1963)
317) Stackelberg, M.v., Stracke, W.: Z Elektrochem. *53*, 118 (1949)
318) Elving, Ph.J., Rosenthal, J, Martin, A.J.: J. Am. Chem. Soc. *77*, 5218 (1955);
318a) Eberson, L.: Acidity and hydrogen bonding of carboxyl groups. In: The chemistry of carboᵛylic acids and esters, chapter 6; Patai, S., ed., 1970
319) Doyle, A.M., Pedler, A.E., Tatlow, I.C.: J. Chem. Soc. C *1968*, 2740
320) Arapakos, P.G., Scott, M.K.: Tetrahedron Letters *1968*, 1975
321) Zavada, J., Krupicka, J., Sicher, J.: Collection Czech. Chem. Commun. *28*, 1664 (1963)
322) Covitz, F.H.: J. Am. Chem. Soc. *89*, 5403 (1967)
323) Gilch, H.G.: J. Polymer Sci. A *1*, *4*, 1351 (1966)
324) Eberson, L.: Electrochim. Acta *12*, 1473 (1967);
324a) Leermakers, P.A., Weis, L.D., Thomas, H.T.: J. Am. Chem. Soc. *87*, 4403 (1965)
325) Svadskovskaya, G.E., Voitkevich, S.A.: Russ. Chem. Rev. (English Transl.) *29*, 161 (1960)
326) Weedon, B.C.L.: Advan. Org. Chem. *1*, 1 (1960
327) Conway, B.E., Vijh, A.K.: J. Org. Chem. *31*, 4283 (1966)
327a) Garwood, R.F., Scott, C.J., Weedon, B.C.L.: Chem. Commun. *1*, 14 (1965)
328) Eberson, L.: Acta Chem. Scand. *17*, 1196 (1963)
328a) Rand, L., Moher, A.F.: J. Org. Chem. *30*, 3885 (1965)
329) Eberson, L., Nilson, S.: unpublished
330) —, Sandberg, B.: Acta Chem. Scand. *20*, 739 (1966)
331) Conway, B.E., Dzieciuch, M.: Can. J. Chem. *41*, 21, 38, 55 (1963). — Conway, B.E., Vijh, A.K.: Z. Anal. Chem. *224*, 149, 160 (1967)
332) Fleischmann, M.F., Mansfield, J.R., Wynne-Jones, W.F.K.: J. Electroanal. Chem. *10*, 511, 522 (1965). — Fleischmann, M., Goodridge, F.: Discussions Faraday Soc. *45*, 254 (1968). — Hickling, A., Wilkins, R,: Discussions Faraday Soc. *45*, 261 (1968)
333) Reichenbacher, P., Liu, M.Y.C., Skell, P.: J. Am. Chem. Soc. *90*, 1816 (1968)
334) Eberson, L.: J. Am. Chem. Soc. *91*, 2402 (1969)
335) Wallis, E.S., Adams, F.H.: J. Am. Chem. Soc. *55*, 3838 (1933). — Smith, W.B., Gilde, H.: J. Am. Chem. Soc. *83*, 1355 (1961)
336) Kazakova, L.I., Fioshin, M.Y., Avtruskaya, I.A.: Zh.Prikl. Khim. *41*, 1326 (1968); Chem. Abstr. *69*, 64097 u (1968)
337) Kazuhiro, M., Katsuya, M.: Nippon Kagaku Zasshi *89*, 196 (1968); Chem. Abstr. *68*, 110809 g (1968)
338) Sokolov, S.V., Levin, A.J., Chechina, O.N.: Zh. Obshch. Khim. *35*, 1718 (1965); Chem. Abstr. *64*, 1943 a (1966)
339) Wohl, A., Schweitzer, H.: Chem. Ber. *39*, 890 (1906)
340) Fichter, F., Lurie, S.: Helv. Chim. Acta *16*, 885 (1933)
341) Offe, H.A.: Z. Naturforsch. *2b*, 182 (1947)
342) Rand, L., Rao, C.S.: J. Org. Chem. *33*, 2704 (1968)
343) Serck-Hanssen, K., Ställberg-Stenhagen, S., Stenhagen, E.: Arkiv Kemi *5*, 203, (1953)

References

344) Haufe, J., Beck, F.: Chem.-Ing.-Tech. *42*, 170 (1970)
345) Woolford, R.G., Lin, W.S.: Can. J. Chem. *44*, 2783 (1966)
346) Binns, T.D., Brettle, R., Cox, G.B.: J. Chem. Soc. C *1968*, 584
347) Vellturo, A.F., Griffin, G.W.: J. Org. Chem. *31*, 2241 (1966)
348) Linstead, R.P., Lunt, J.C., Weedon, B.C.L.: J. Chem. Soc. *1950*, 3331;
348a) Eberson, L.: J. Org. Chem. *27*, 3706 (1962)
349) Milburn, A.H., Truter, E.V.: J. Chem. Soc. *1954*, 3344
350) Baker, B.W., Linstead, R.P., Weedon, B.C.L.: J. Chem. Soc. *1955*, 2218
351) Brettle, R., Parkin, J.G.: J. Chem. Soc. C *1967*, 1352
352) Okubo, T., Tsutsumi, S.: Bull. Soc. Chem. Japan *37*, 1794 (1964)
353) – –, Nippon Kagaku Zassi *87*, 449 (1966)
354) Adickes, F., Brunnert, W., Lücker, O.: J. Prakt. Chem. *130*, 163 (1931)
355) Ulpiani, C., Rodano, G.A.: Atti Accad. Naze. Lincei *14*⌊II⌋604 (1905)
356) Bahner, C.T.: U.S.Pat. 2, 485, 803 (1949), Chem. Abstr. *44*, 2876 b (1950)
357) –, Ind. Eng. Chem. *44*, 317 (1952)
358) Johnston, K.M.: Tetrahedron Letters *1967*, 837 ;
358a) Fichter, F., Christen, A.: Helv. Chim. Acta *8*, 332 (1925)
359) Bobitt, J.M., Stock, J.T., Marchand, A., Weisgraber, K.H.: Chem. Ind. (London) *1966*, 2127
360) Lebedeva, A.J.: Dokl. Akad. Nauk. SSSR *42*, 71 (1944)
361) Bunge, N.: Chem. Ber. *3*, 911 (1870)
362) –, *3*, 295 (1870)
363) Schall, C., Kraszler, S.: Z. Elektrochem. *5*, 225 (1899)
364) Evans, W.V., Braithwaite, D., Field, E.F.: J. Am. Chem. Soc. *62*, 534 (1940)
365) Morgat, J.L., Pallaud, R.: Compt. Rend. *260*, 5579 (1965)
366) – –, *260*, 574 (1965) ;
366a) Evans, W.V., Braithwaite, D.: J. Am. Chem. Soc. *61*, 898 (1939). – Evans, W.V., Field, E.: J. Am. Chem. Soc. *58*, 720 (1936); *58*, 2284 (1936). – Evans, W.V., Lee, F.H.: J. Am. Chem. Soc. *56*, 654 (1934). – Dessy, R.E., Psarras, Th.: J. Am. Chem. Soc. *88*, 5132 (1966);
366b) Schöllkopf, U.: Herstellung und Umwandlung lithiumorganischer Verbindungen, Houben-Weyl, Methoden der organischen Chemie, Bd. 13, p. 3,87, Stuttgart: G. Thieme. – Mallan, J.M., Bebb, R.L.: Chem. Rev. *69*, 693 (1969). – Schlosser, M.: Angew. Chem. *76*, 124 (1964). – Krause, E., Grosse, A.v.: Die Chemie der metallorganischen Verbindungen, Wiesbaden: 1937, M. Ständig oHG
366c) Beck, F., Guthke, H.: Chem. Ing.-Tech. *41*, 943 (1969)
367) Cooks, R.G., Williams, D.H., Johnston, K.M., Stride, J.D.: J. Chem. Soc. C *1968*, 2199. – Johnston, K.M., Stride, J.D.: Chem. Commun. *1966*, 325
368) Neth. App. P 6, 508, 611; see Chem. Abstr. *67*, 17392 (1967)
369) Osa, T., Yildiz, A., Kuwana, T.: J. Am. Chem. Soc. *91*, 3994 (1969)
370) Wisdom, N.E., Jr.: Extended abstracts, Meeting of the Electrochemical Society: New York, May 4 – 9, 1969, p. 338
371) Fichter, F., Rinderspacher, M.: Helv. Chim. Acta *10*, 40 (1927)
372) –, Herzbein, S.: Helv. Chim. Acta *11*, 1264 (1928)
373) Erdtmann, G.H.: Proc. Roy. Soc. (London) A *143*, 191 (1934)
374) a) Friend, K.E., Ohnesorge, W.E.: J. Org. Chem. *28*, 2435 (1963);
b) Parker, V.D.: Chem. Commun. *1969*, 1131
375) Ohnesorge, W.E., Stuart, J.D.: Extended abstracts, Meeting of the Electrochemical Society, New York, May 4 – 9, 1969, p. 305
376) Wawzonek, S., Mc Intyre, T.W.: J. Electrochem. Soc. *114*, 1025 (1967)
377) Bacon, J., Adams, R,N.: J. Am. Chem. Soc. *90*, 6596 (1968)
378) Cauquis, G., Badoz-Lambling, J., Billon, J.P.: Bull. Soc. Chim. France *1965*, 1433
379) Galus, Z., White, R.M., Rowland, F.S., Adams, R.N.: J. Am. Chem. Soc. *84*, 2065 (1962)
380) Seo, E.T., Nelson, R.F., Fritsch, J.M., Marcoux, L.S., Leedy, D.W., Adams, R.N.: J. Am. Chem. Soc. *88*, 3498 (1966)
381) Ambrose, J.F., Nelsen, R.F.: J. Electrochem. Soc. *115*, 1159 (1968)

382) Fichter, F., Ackermann, F.: Helv. Chim. Acta 2, 583 (1919)
383) German Pat. 85, 390 (1895), C. I. 1248 (1896)
384) Schäfer, H., Steckhan, E.: Tetrahedron Letters, 3835 (1970)
385) Zhurinov, M.Zh., Fioshin, M.Y., Mirkind, L.A.: Elektrokhymyia 1969, 1257
386) Fritsch, J.M., Weingarten, H.: J.Am. Chem. Soc. 90, 793 (1968); 92, 4038 (1970)
387) Jugelt, W., Pragst, F.: Angew. Chem. 80, 280 (1968); Tetrahedron 24, 5123 (1968)
388) Baizer, M.M., Petrovich, J.P.: In: Progress in Physical Organic Chemistry, Vol. 7, p. 189; ed. by Streitwieser, A., Jr., Taft, R.W.
389) Sekine, T., Yamura, A., Sugino, K.: J. Electrochem. Soc. 112, 439 (1965)
390) Tomilov, A.P., Kryeukova, E.V., Klimov, V.A., Brage, I.N.: Soviet Electrochem. 3, 1352 (1967)
391) Arai, T.: Denki Kagaku 30, 31 (1962); Chem. Abstr. 62, 15760e (1965)
392) —, Denki Kagaku 30, 175 (1962); Chem. Abstr. 62, 15760g (1965)
393) —, Denki Kagaku 30, 235 (1963); Chem. Abstr. 62, 15761a (1965)
394) Tomilov, A.P., Kalitina, M.I.: Zh. Prikl. Khim. 38, 1574 (1965); Chem. Abstr. 63, 11343e (1965)
395) Fioshin, M.Y., Avrutskaya, I.A., Gerasimova, L.E., Kabnova, S.D.: Chem. Abstr. 71, 30158 (1969); Khim. Pharm. Zh. 3, 19
396) Allen, M.J.: J. Org. Chem. 15, 435 (1950)
397) —, Corvin, A.H.: J. Am. Chem. Soc. 72, 114 (1952)
398) Leonard, N.J., Swann, S. Jr., Fuller, G.: J. Am. Chem. Soc. 75, 5127 (1953)
399) Allen, M.J.: J. Am. Chem. Soc. 72, 3797 (1950)
400) Albert, W.C., Lowy, A.: Trans. Electrochem. Soc. 75, 367 (1939)
401) Allen, M.J., Fearn, J.F., Levine, H.A.: J. Chem. Soc. 1952, 2220
402) Pasternak, R.: Helv. Chim. Acta 31, 753 (1948)
403) Juday, R.E.: Symposium on the Synthetic and Mechanistic aspects of Organic Electrochemistry, U.S. Army Research Office, Durham, N.C., U.S.A., October 14-16, 1968, pp. 211
404) Stocker, J.H., Jenevein, R.M.: J. Org. Chem. 33, 294 (1968); 33, 2145 (1968)
405) Curphy, T.J., Amelotti, W., Layloff, T.P., McCartney, R.L., Williams, J.H.: J. Am. Chem. Soc. 91, 2817 (1969)
406) Curphey, T.J., Amelotti, C.W., Layloff, T.P., McCartney, R.L., Williams, J.H.: Extended abstracts, Meeting of the Electrochemical Society, New York, May 4-9, 1969, p. 344
407) Horner, L., Skaleta, D.H.: Tetrahedron Letters 1970, 1103
408) Pearl, I.A.: J. Am. Chem. Soc. 74, 4260 (1952)
409) Allen, M.J., Cohen, H.: J. Electrochem. Soc. 106, 451 (1959)
410) Gouwley, R.N., Grimshaw, J.: J. Chem. Soc. C 1968, 2388
411) Wiemann, J.: Bull. Soc. Chim. France 1964, 2545
412) —, Bouguerra, M.L.: Ann. Chim. (Paris) 3, 215 (1967)
413) Simonet, J.: Compt. Rend. Ser. C 267, 1548 (1968)
414) Wiemann, J., Bouguerra, M.L.: Ann. Chim. (Paris) 2, 35 (1967)
415) Touboul, E., Weisbuch, F., Wiemann, J.: Compt. Rend. Ser. C 268, 1170 (1969)
416) Ivcher, T.S., Zilberman, E.N., Perepletchikova, E.: Chem. Abstr. 70, 111 113d (1969)
417) Mazzenga, A., Lomnitz, D., Villegas, J., Polowczyk, C.: Tetrahedron Letters 1969, 1665
418) Bowers, K.W., Giese, R.W., Grimshaw, J., House, H.O., Kolodny, N.H., Kronberger, K., Roe, D.K.: J. Am. Chem. Soc. 92, 2783 (1970)
419) Baizer, M.M., Anderson, J.D.: J. Electrochem. Soc. 111, 223 (1964)
420) —, Petrovich, J.P.: J. Electrochem. Soc. 114, 1023 (1967)
421) Bouguerra, M.L., Wiemann, J.: Compt. Rend. Ser. C 263, 782 (1966)
422) Baizer, M., Ort, M.R.: J. Org. Chem. 31, 1646 (1966)
423) —, Anderson, J.D., Wagenknecht, J.H., Ort, M.R., Petrovich, J.P.: Electrochim. Acta 12, 1377 (1967)
424) —, J. Electrochem. Soc. 111, 215 (1964)
425) U.S. Pat. 3,193,480/81 (1961) Fa. Monsanto; see Prescott, J.H.: Chem. Eng. 72, 238 (1965)

References

426) Frz. Pat. 1,548,304, Chem. Abstr. *71*, 80743y; Frz. Pat. 1491516, Chem. Abstr. *69*, 7875x (1968); Frz. Pat. 1,503,244, Chem. Abstr. *69*, 76685x (1968); Brit. Pat. 1,117,677, Chem. Abstr. *69*, 43, 483b (1968); see also: Beck, F.: Chem.-Ing.-Tech. *42*, 153 (1970)

427) Belg. Pat. 695 154 (UCB, 1966); Matsuda, F.: Tetrahedron Letters *1966*, 6193

428) Baizer, M.M., Anderson, J.D.: J. Electrochem. Soc. *111*, 226 (1964)

429) Anderson, J.D., Baizer, M.M., Prill, E.J.: J. Org. Chem. *30*, 1645 (1965)

430) Baizer, M.M.: U.S. Pat. 3,249,521, Chem. Abstr. *65*, 7217g (1966)

431) —, Anderson, J.D.: J. Org. Chem. *30*, 1348 (1965)

431a) Brand, K., Krücke-Amelung, D.: Chem. Ber. *72*, 1029 (1939)

432) Anderson, J.D., Petrovich, J.P., Baizer, M.M.: Advan. Org. Chem., Interscience 1969, Vol. 6, p. 257

433) Anderson, J.D., Baizer, M.M., Petrovich, J.P.: J. Org. Chem. *31*, 3890 (1966). — Anderson, J.D., Baizer, M.M.: Tetrahedron Letters *1966*, 511. — Pat. Monsanto BP 1,127,368; Chem. Abstr. *70*, 37839b (1969)

434) McClemens, D.J., Garrison, A.K., Underwood, A.L.: J. Org. Chem. *34*, 1867 (1969)
a) Brand, K., Krücke-Amelung, D.: Chem. Ber. *72*, 1029 (1939)

435) Jones, G.C., Ledford, T.H.: Tetrahedron Letters *1967*, 615

436) Baizer, M.M.: Tetrahedron Letters *1963*, 973

437) Beck, F.: Chem.-Ing.-Tech. *37*, 607 (1965)

438) Tomilov, A.P., Fecktistov, L.G.: Soviet Electrochem. *1*, 1165 (1965)

439) Gillet, J.: Bull. Soc. Chim. France *1968*, 2919

440) Asahara, T., Seno, M., Arai, T.: Bull. Chem. Soc. Japan *42*, 1316 (1969)

441) Lazarov, S., Trifunov, A., Vitanov, T.: Z. Physik. Chem. (Leipzig) *226*, 221 (1964)

442) Figeys, M.: Tetrahedron Letters *1967*, 2237

443) Petrovich, J.P., Baizer, M.M., Ort, M.R.: J. Electrochem. Soc. *116*, 743 (1969)

444) —, J. Electrochem. Soc. *116*, 749 (1969)

445) —, Anderson, J.D., Baizer, M.M.: J. Org. Chem. *31*, 3897 (1966)

446) Levine, H.A., Allen, M.J.: J. Chem. Soc. *1952*, 254

447) Allen, M.J., Siragusa, J.A., Pierson, W.: J. Chem. Soc. *1960*, 1045

448) Sugino, K., Nonaka, T.: Electrochim. Acta *13*, 613 (1968). — Sugino, K.: J. Electrochem. Soc. *112*, 1241 (1965)

449) Klynev, B.L., Tomilov, A.P.: Zh. Obshch. Khim. *39*, 470 (1969); Chem. Abstr. *70*, 120479 (1969)

450) Brown, O.R., Lister, K.: Discussions Faraday Soc. *45*, 106 (1968)

451) Pallaud, R., Nicolas, M.: Compt. Rend. Ser. C *267*, 1834 (1968)

452) Wagenknecht, J.H., Baizer, M.M.: J. Org. Chem. *31*, 3885 (1966)

453) Sugino, K., Nonacka, T.: J. Electrochem. Soc. *116*, 615 (1969). — Ferles, M., Vanka, M., Silhankovic, A.: Collection Czech. Chem. Commun. *1969*, 2108

454) Baizer, M.M.: J. Org. Chem. *29*, 1670 (1964). — Baizer, M.M.: U.S. Pat. 3,193,476 Chem. Abstr. *63*, 13092b (1964)

455) —,: J. Org. Chem. *31*, 3847 (1966); U.S. Pat. 3,440,154, Chem. Abstr. *71*, 70351e (1969)

456) —,: U.S. Pat. 3,438,877, Chem. Abstr. *71*, 49534 (1969)

457) —, Anderson, J.D.: J. Org. Chem. *30*, 1351 (1965)

458) Miller, L.L., Hoffmann, A.K.: J. Am. Chem. Soc. *89*, 593 (1967)

459) —, Kaufmann, D.A., Kujawa, E.P.: The Electrochemical Society, Spring Meeting 1969, New York, Extended abstracts No. 131, p. 314

460) O'Donnell, J.F., Mann, Ch.K.: J. Electroanal. Chem. *13*, 157 (1967)

461) Mizuno, S.: J. Electrochem. Soc. Japan *29*, 27, 31, 33, 112 (1961); Chem. Abstr. *62*, 15761b (1965)

462) —,: Denki Kagaku *29*, 106 (1961); Chem. Abstr. *62*, 3667b (1965)

463) Barnes, K.K., Mann, Ch.K.: J. Org. Chem. *32*, 1474 (1967)

464) Smith, P.J., Mann, Ch.K.: J. Org. Chem. *34*, 1821 (1969)

465) Cottrell, P.T., Mann, Ch.K.: J. Electrochem. Soc. *116*, 1499 (1969)

466) Parker, V.D.: Chem. Commun. *1969*, 610

467) Kemula, W., Grabowski, Z.R., Kalinowski, M.K.: J. Am. Chem. Soc. *91*, 6863 (1969). – Michielli, R.F., Elving, Ph.J.: J. Am. Chem. Soc. *91*, 6864 (1969)

468) Maruyama, K., Murakami, K.: Bull. Chem. Soc. Japan *1968*, 41, 1401

469) Oyama, M., Ohno, M.: Tetrahedron Letters *1966*, 5201

470) Elving, Ph.J.: Record Chem. Progr. (Kresge-Hooker Sci.Lib.) *14*, 99 (1953)

471) Hush, N.S., Segal, G.A.: Discussions Faraday Soc. *45*, 23 (1968). – Zhdanov, S.I.: Electrochim. Acta *10*, 657 (1965)

472) Calvert, J.G., Pitts, J.N. Jr.: Photochemistry, p. 522. New York, J. Wiley & Sons, Inc., 1966

473) Sease, J.W., Chang, P., Groth, J.L.: J. Am. Chem. Soc. *86*, 3154 (1964)

474) Rosenthal, I., Albright, C.A., Elving, Ph.J.: J. Electrochem. Soc. *99*, 227 (1952). – Elving, Ph.J., Markowitz, J.M.: J. Electrochem. Soc. *101*, 195 (1954)

475) Streitwieser, A..Jr., Perrin, Ch.: J. Am. Chem. Soc. *86*, 4938 (1964)

476) Applequist, D.E., Kaplan, L.: J. Am. Chem. Soc. *87*, 2194 (1965)

477) Sease, J.W., Reed, R.C.: The Electrochemical Society, Spring Meeting 1969, New York, Extended abstract No. 134, p. 328

478) Lambert, F.L.: J. Org. Chem. *31*, 4184 (1966)

479) –, Albert, A.H., Hardy, J.P.: J. Am. Chem. Soc. *86*, 3155 (1964)

480) Krupcicka, J., Zavadand, J., Sicher, J.: Collection Czech. Chem. Commun. *30*, 3570, (1965). – Zuman, P.: Talanta *12*, 1337 (1965)

481) Feoktistov, L.G., Zhdanov, S.I.: Electrochim. Acta *10*, 657 (1965). – Tomilov, A.P., Smirnov, Y.D., Goldin, M.M.: Sov. Electrochem. *1*, 791 (1965); Chem. Abstr. *63*, 16196e (1965). – Tomilov, A.P., Smirnov, Y.D., Varshavskii, S.L.: J. Gen. Chem. USSR *35*, 390 (1965). – Tomilov, A.P.: J. Chem. Gen. USSR *38*, 218 (1968)

482) Kaabak, L.V., Tomilov, A.P., Vars, S.L., Kabachnik, I.: J. Org. Chem. USSR *3*, 1 (1967)

483) Plump, R.E., Hammett, L.P.: Trans. Electrochem. Soc. *73*, 523 (1938)

484) Lawless, J.G., Bartak, D.E., Hawley, M.D.: J. Am. Chem. Soc. *91*, 7121 (1969)

485) Grimshaw, J., Ramsey, J.S.: J. Chem. Soc. B *1968*, 60

486) Rogers, L.B., Diefenderfer, A.N.: J. Electrochem. Soc. *114*, 942 (1967)

487) Eberson, L.: Acta Chem. Scand. *22*, 3045 (1968)

488) Azoo, J.A., Coll, F.J., Grimshaw, J.: J. Chem. Soc. C *1969*, 2521

489) Mann, C.K., Webb, J.L., Walborsky, H.M.: Tetrahedron Letters *1966*, 2249

490) Elving, Ph.J., Rosenthal, I., Hayes, J.R., Martin, H.J.: Anal. Chem. *33*, 330 (1961)

491) Miller, L.L., Riekena, E.: J. Org. Chem. *34*, 3359 (1969)

492) Annino, R., Erickson, R.E., Michalovic, J., McKay, B.: J. Am. Chem. Soc. *88*, 4424 (1966)

493) Czochralska, B.: Chem. Phys. Letters *1*, 239 (1967)

494) Fry, A.J., Mitnick, M., Reed, R.G.: J. Org. Chem. *35*, 1232 (1970)

495) Nagao, M., Sato, N., Akashi, T., Yoshida, T.: J. Am. Chem. Soc. *88*, 3448 (1966)

496) Stocker, J.H., Jenevein, R.M.: Chem. Commun. *1968*, 934

497) Carrahar, P., Drakesmith, F.G.: Chem. Commun. *1968*, 1562

498) Coleman, J.P., Gilde, H.G., Utley, J.H.P., Weedon, B.C.L.: Chem. Commun. *1970*, 738

499) Lund, H.: Acta Chem. Scand. *13*, 192 (1959)

500) Horner, L., Röder, H.: Chem. Ber. *101*, 4197 (1968)

501) Wrobel, J.T., Bien, A.S., Pazdro, K.M.: Chem. Ind. (London) *1966*, 1760. – Wrobel, J.T., Krawczyk, A.R.: Chem. Ind. (London) *1969*, 656

502) Finkelstein, M., Petersen, R.C., Ross, S.D.: J. Am. Chem. Soc. *81*, 2361 (1959)

503) Ross, S.D., Finkelstein, M., Petersen, R.C.: J. Am. Chem. Soc. *82*, 1582 (1960)

504) Dubois, J.E., Monvernay, A., Lacaze, P.C.: Electrochim. Acta *15*, 315 (1970)

505) Horner, L., Mentrup, A.: Liebigs Ann. Chem. *646*, 49 (1961)

506) –, –,: Liebigs Ann. Chem. *646*, 65 (1961)

507) –, Winkler, H., Rapp, A., Mentrup, A., Hoffmann, H., Beck, P.: Tetrahedron Letters *1961*, 161

508) Bestmann, H., Vilsmaier, E., Graf, G.: Liebigs Ann. Chem. *704*, 109 (1967)

References

509) Horner, L., Röttger, F., Fuchs, H.: Chem. Ber. *96*, 3141 (1963). – Horner, L., Haufe, J.: Chem. Ber. *101*, 2903 (1968)
510) –, Fuchs, H.: Tetrahedron Letters *1962*, 203
511) –, Neumann, H.: Chem. Ber. *98*, 3462 (1965)
512) –, Singer, R.: Liebigs Ann. Chem. *723*, 1 (1969
513) Pat. Neth. Appl. 6, 602, 261; Chem. Abstr. *66*, 38237 (1967)
514) Horner, L., Singer, R.: Chem. Ber. *101*, 3329 (1968)
515) Yousefazedeh, P., Mann, C.K.: J. Org. Chem. *33*, 2716 (1968)
516) Horner, L., Neumann, H.: Chem. Ber. *98*, 1715 (1965)
517) Ives, D.A.J.: Can. J. Chem. *47*, 3697 (1969)
518) Lund, H.: Acta Chem. Scand. *14*, 1927 (1960)
519) Leonard, N.J., Swann, S., Jr., Figueras, J.: J. Am. Chem. Soc. *74*, 4620 (1952)
520) –, –, Mottus, E.H.: J. Am. Chem. Soc. *74*, 6251 (1952)
521) Lamm, B. , Samuelson, B.: Acta Chem. Scand. *23*, 691 (1969)
522) Wawzonek, S., Fredrickson, J.D.: J. Electrochem. Soc. *106*, 325 (1959)
523) Arapakos, P.G., Scott, M.K.: Tetrahedron Letters *1968*, 1975
524) Hoffmann, A.K., Hodgson, W.G., Maricle, D.L., Jura, W.H.: J. Am. Chem. Soc. *86*, 631 (1964)
525) Pysh, E.S., Yang, N.C.: J. Am. Chem. Soc. *85*, 2124 (1963)
526) Loveland, J.W., Dimeler, G.R.: Anal. Chem. *33*, 1196 (1961)
527) Neikam, W.C., Dimeler, G.R., Desmond, M.M.: J. Electrochem. Soc. *111*, 1190 (1964)
528) Neikam, W.C., Desmond, M.M.: J. Am. Chem. Soc. *86*, 4811 (1964)
529) Steitwieser, A., Jr.: Molecular orbital theory for Organic Chemists. New York: J.Wiley & Sons, 1961
530) Heilbronner, E., Bock, H.: Das HMO Modell und seine Anwendung, I. Weinheim: Verlag Chemie, 1968
531) Phelps, J., Santhanam, K.S.V., Bard, A.J.: J. Am. Chem. Soc. *89*, 1752 (1967)
532) Marcoux, L.S., Fritsch, J.M., Adams, R.N.: J. Am. Chem. Soc. *89*, 5766 (1967)
533) Peover, M.E., White, B.S.: J. Electroanal. Chem. *13*, 93 (1967)
534) Fleischmann, M., Pletcher, D.: J. Electroanal. Chem. *25*, 449 (1970)
535) Billon, J.P. Cauquis, G., Raison, J., Thibaud, Y.: Bull. Soc. Chim. France *1967*, 199
536) Tonnard, F.: Compt. Rend. *260*, 2793 (1965)
537) Cauquis, G., Genies, M., Lemaire, H., Rassat, A., Ravet, J.P.: J. Chem. Phys. *47*, 4642 (1967)
538) Zweig, A., Hodgson, W.G., Jura, W.H.: J. Am. Chem. Soc. *86*, 4124 (1964)
539) –, Maurer, A.H., Roberts, B.G.: J. Org. Chem. *32*, 1322 (1967)
540) –, Hoffmann, A.K.: J. Org. Chem. *30*, 3997 (1965)
541) –, Hodgson, W.G., Jura, W.H., Maricle, D.L.: Tetrahedron Letters *1963*, 1821
542) Nelson, R.F., Adams, R.N.: J. Am. Chem. Soc. *90*, 3925 (1968)
543) Cauquis, G., Genies, M.: Bull. Soc. Chim. France *1967*, 3220
544) Adams, R.N., Piette, L.H., Ludwig, P.: Anal. Chem. *34*, 916 (1962)
545) Melchior, M.T., Maki, A.H.: J. Chem. Phys. *34*, 471 (1961)
546) Smejtek, P., Honzl, J., Metalova, V.: Collection Czech. Chem. Commun. *30*, 3875 (1965)
547) Nelson, R.F., Leedy, D.W., Seo, E.T., Adams, R.N.: Z. Anal. Chem. *224*, 185 (1967)
548) Zweig, A., Metzler, G., Maurer, A.H., Roberts, B.G.: J. Am. Chem. Soc. *89*, 4091 (1967)
549) Billon, J.P., Cauquis, G., Coubrisson, J.: J. Chim. Phys. *61*, 374 (1964) and earlier papers
550) Hünig, S., Schlaf, H., Kießlich, G., Scheutzow, D.: Tetrahedron Letters *1969*, 2271
551) Hercules, D.M.: Accounts Chem. Res. *2*, 301 (1969)
552) Kuwana, Th.: Advan. Electroanal. Chem. Vol. 1,p.197; Bard, A.J., ed. M. Dekker 1966
553) Zweig, A., Metzler, G., Maurer, A., Roberts, B.G.: J. Am. Chem. Soc. *88*, 2864 (1966)
554) Hercules, D.M.: Science *145*, 808 (1964). – Hercules, D.M., Lansbury, R.C., Roe, D.K.: J. Am. Chem. Soc. *88*, 4578 (1966)

555) Visco, R.E., Chandross, E.A.: J. Am. Chem. Soc. *86*, 5350 (1964)
556) Cruser, S.A., Bard, A.J.: J. Am. Chem. Soc. *91*, 267 (1969). – Malby, J.T., Prater, K.B., Bard, A.J.: J. Phys. Chem. *72*, 4348 (1968)
557) Faulkner, L.R., Bard, A.J.: J. Am. Chem. Soc. *91*, 209 (1969)
558) Visco, R.E., Chandros, E.A.: Electrochim. Acta *13*, 1187 (1968)
559) Chang, J., Hercules, D.M., Roe, D.K.: Electrochim. Acta *13*, 1197 (1968)
560) Zweig, A., Maricle, D.L., Brinen, J.S., Maurer, A.H.: J. Am. Chem. Soc. *89*, 473 (1967)
561) Vodzinski, V.J., Vasileva, A.A., Atakumov, G., Korshunov, J.A.: Elektrokhimiya *1968*, 1492; Chem. Abstr. *70*, 73531 d (1969)
562) Steuber, F.W., Dimroth, K.: Chem. Ber. *99*, 258 (1966)
563) Tomilov, A.P., Smirnov, Y.D., Videiko, A.F.: Elektrochimiya *2*, 603 (1966); Chem. Abstr. *65*, 3343 e (1966)
564) Kuwata, K., Geske, D.H.: J. Am. Chem. Soc. *86*, 2101 (1964)
565) Mc Clelland, B.J.: Chem. Rev. *64*, 301 (1964). – Kaiser, E.T.: Radical Ions. New York: Interscience 1968;
565a) Kastening, B.: Chem.-Ing.-Tech. *42*, 190 (1970)
566) Wawzonek, S., Laitinen, H.A.: J. Am. Chem. Soc. *64*, 2365 (1942). – Hoijtink, G.J., Schooten, J.v.: Rec. Trav. Chim. *72*, 691, 903 (1953). – Bergmann, I.: Trans. Faraday Soc. *52*, 690 (1956);
566a) Pryor, W.A.: Free Radicals, p. 242: W.A. Benjamin 1966
566b) Cauquis, G.: Bull. Soc. Chim. France 1968, 1618
567) Hoijtink, G.L: Rec. Trav. Chim. *73*, 819 (1954)
567a) Dohrmann, J.K., Vetter, K.J.: J. Electroanal. Chem. *20*, 23 (1969)
568) Dietz, R., Peover, M.E.: Discussions Faraday Soc. *45*, 154 (1968). – Dietz, R., Peover, M.E.: Trans. Faraday Soc. *62*, 3535 (1966)
569) Szwarc, M.: Accounts Chem. Res. *2*, 87 (1969)
570) –, Asami, R.: J. Am. Chem. Soc. *84*, 2269 (1964)
571) Hoijtink, G.J.: Rec. Trav. Chim. *76*, 869 (1957)
572) Umemoto, K.: Bull. Soc. Chem. Japan *40*, 1058 (1967)
573) Santhanam, K.S.V., Bard, A.J.: J. Am. Chem. Soc. *88*, 2669 (1966)
574) Allendoerfer, R.D., Rieger, Ph.H.: J. Am. Chem. Soc. *87*, 2336 (1965)
575) Katz, T.J., Talcott, C.: J. Am. Chem. Soc. *88*, 4732 (1966)
576) Tolle, W.M., Moore, D.W.: J. Chem. Phys. *46*, 2102 (1967)
577) Levy, D.H., Myers, R.J.: J. Chem. Phys. *44*, 4177 (1966)
578) Geske, D.H., Balch, A.L.: J. Phys. Chem. *68*, 3423 (1964)
579) Dehl, R., Fraenkel, G.K.: J. Chem. Phys. *39*, 1793 (1963)
580) Johnson, C.S., Chang, R.J.: J. Chem. Phys. *43*, 3183 (1965)
581) Bauld, N.L., Stevensson, G.R.: J. Am. Chem. Soc. *91*, 3675 (1969)
582) Geske, D.H., Maki, A.H.: J. Am. Chem. Soc. *82*, 2671 (1960)
583) Maki, A.H., Geske, D.H.: J. Am. Chem. Soc. *83*, 1852 (1961)
584) Harriman, J.E., Maki, H.: J. Chem. Phys. *39*, 778 (1963)
585) Kazakova, V.M., Makarov, I.G., Kurek, M.E., Cernyshey, E.A.: Zh. Strukt. Khim. *9*, 525 (1968); Chem. Abstr. *69*, 105703 j (1968)
586) Allendoerfer, R.D., Rieger, Ph.H.: J. Am. Chem. Soc. *88*, 3711 (1966)
587) Chambers, R.Q., III, Adams, R.N.: J. Electroanal. Chem. *9*, 400 (1965)
588) Kemula, W., Sioda, R.: J. Electroanal. Chem. *7*, 233 (1964)
589) Greig, W.N., Rogers, J.W.: J. Am. Chem. Soc. *91*, 5495 (1969)
590) Piette, L.H., Ludwig, P., Adams, R.N.: J. Am. Chem. Soc. *84*, 4212 (1962). – Sayo, H., Masui, M.: Tetrahedron *24*, 5075 (1968)
591) Rieger, Ph.H., Bernal, I., Reinmuth, W.H., Fraenkel, G.K.: J. Am. Chem. Soc. *85*, 683 (1963)
592) –, –, Fraenkel, G.K.: J. Am. Chem. Soc. *83*, 3918 (1961)
593) Dessy, R.E., Kleiner, M., Cohen, S.C.: J. Am. Chem. Soc. *91*, 6800 (1969)
595) Horner, L., Singer, R.: Tetrahedron Letters *1969*, 1545
596) Brodsky, A.E., Gordienko, L.L., Degtiarev, L.S.: Electrochim. Acta *13*, 1095 (1968)
597) Schwarz, W.M., Kosower, M., Shain, I.: J. Am. Chem. Soc. *83*, 3164 (1961)

References

598) Volz, H., Lotsch, W.: Tetrahedron Letters *1969*, 2275
599) Kothe, G., Sümmermann, W., Baumgärtel, H., Zimmermann, H.: Tetrahedron Letters *1969*, 2186
600) James, M.I., Plesch, P.H.: Chem. Commun. *1967*, 508
601) Hausser, K.H., Häbich, A., Franzen, V.: Z. Naturforsch. *16 A*, 836 (1961)
602) Funt, L., Gray, D.G.: The Electrochemical Society, Spring Meeting 1969, New York, Abstr. 135, p. 330
609) Balch, A.L.: J. Am. Chem. Soc. *91*, 1948 (1969) and earlier papers. − Olson, D.C., Mayweg, V.P., Schrauzer, J.N.: J. Am. Chem. Soc. *88*, 4876 (1966)
610) Marlett, E.M.: Ann. N.Y. Acad. Sci. *125*, 12 (1965)
611) Chambers, J.Q., Norman, A.D., Bickell, M.R., Cadle, S.H.: J. Am. Chem. Soc. *90*, 6056 (1968). − Wiersema, R.J., Middaugh, R.L.: Inorg. Chem. *8*, 2074 (1969)
612) Dessy, R.E., Charkoudian, J.C., Abeles, T.P., Rheingold, A.L.: submitted to publication in J. Am. Chem. Soc.
613) Ziegler, K., Steudel, O.W.: Liebigs Ann. Chem. *652*, 1 (1962)
614) −, Lehmkuhl, H.: Angew. Chem. *67*, 424 (1955)
615) −, −, Germ. Pat. 1, 161, 562 (1964); Chem. Abstr. *60*, 11623 (1964)
616) −, Belg. Pat. 617, 628 (1962); Chem. Abstr. *60*, 3008 (1964)
617) US Pat. 3256 161, 3 380 899 (NALCO Chem. CO), see Beck, F.: Chem.-Ing.-Tech. *42*, 153 (1970)
618) Lehmkuhl, H., Schäfer, R., Ziegler, K.: Chem.-Ing.-Tech. *36*, 612 (1964)
619) Chadwick, J.R., Kinsella, E.: J. Organometal. Chem. *4*, 334 (1965)
620) Mottus, E.H., Ort, M.R.: The Electrochemical Society, Spring Meeting New York, 1969, Extended abstract Nr. 132, p. 316
621) Perruzo, V., Plazzogna, G., Tagliavini, G.: J. Organometal. Chem. *18*, 89 (1969)
622) McKeever, L.D., Waack, R.: J. Organometal. Chem. 17, 142 (1969)
623) Dessy, R.E., Kitching, W., Chivers, T.: J. Am. Chem. Soc. *88*, 453 (1966)
624) −, −, Psarras, Th., Salinger, R., Chen, A., Chivers, T.: J. Am. Chem. Soc. *88*, 460 (1966)
625) −, Chivers, T., Kitching, W.: J. Am. Chem. Soc. *88*, 467 (1966)
626) −, Stary, F.E., King, R.B., Waldrop, M.: J. Am. Chem. Soc. *88*, 471 (1966)
627) −, King, R.B., Waldrop, M.: J. Am. Chem. Soc. *88*, 5113 (1966)
628) −, Weissmann, P.M., Pohl, R.L.: J. Am. Chem. Soc. *88*, 5117 (1966)
629) −, Pohl, R.L., King, R.B.: J. Am. Chem. Soc. *88*, 5121 (1966)
630) −, Weissmann, P.M.: J. Am. Chem. Soc. *88*, 5124 (1966)
631) −, −, J. Am. Chem. Soc. *88*, 5129 (1966)
632) Psarras, Th., Dessy, R.E.: J. Am. Chem. Soc. *88*, 5132 (1966)
633) Fleet, B., Jee, R.D.: J. Electroanal. Chem. *25*, 397 (1970)
634) Tsutsumi, S., Yosahida, K.: J. Org. Chem. Soc. *32*, 468 (1967)
635) Breitenbach, J.W., Srna, C.: Pure Appl. Chem. *4*, 245 (1962)
636) Friedlander, H.Z.: Encykl. Polymer Sci. Technol. *5*, 629 (1966)
637) Funt, B.L.: Macromol. Rev. *1*, 35 (1967). − Yamazaki, N.: Fortschr. Hochpolymerenforsch. *6*, 337 (1969)
638) Shapoval, G.S., Skobets, E.M., Markova, N.P.: Dokl. Akad. Nauk SSSR *173* (2), 392 (1967); through Chem. Abstr. *67*, 11894 (1967)
639) Yamazaki, N., Nakahama, S., Kambara, S.: J. Polymer Sci. B *3*, 57 (1965)
640) Smith, W.B., Gilde, H.-G.: J. Am. Chem. Soc. *82*, 659 (1960)
641) Bockris, J.O.M.: J. Electroanal. Chem. *9*, 408 (1965)

Received September 2, 1970

Graebe: Geschichte der organischen Chemie

Erster Band einer Darstellung in zwei Monographien
Von Carl Graebe
Reprint der Erstauflage Berlin 1920. X, 406 Seiten
Subskriptionspreis bis zum 30. 9. 1971. Gebunden DM 48,—
Späterer Ladenpreis: Gebunden DM 56,—

Walden: Geschichte der organischen Chemie seit 1880

Zweiter Band zu Graebe: „Geschichte der organischen Chemie"
Von Paul Walden
Reprint der Erstauflage Berlin 1941. XIV, 946 Seiten
Subskriptionspreis bis zum 30. 9. 1971: Gebunden DM 84,—
Späterer Ladenpreis: Gebunden DM 96,—

Die Nachdrucke sind abhängig von dem zeitgerechten Eingang einer genügenden Anzahl von Bestellungen

Diese Geschichte der organischen Chemie, die bis zum Jahr 1940 reicht, kann angesehen werden als „Handbuch der klassischen Probleme der organischen Chemie".
Band 1 schildert die „Sturm- und Drangperiode und kämpferische Jugendzeit" der wissenschaftlichen organischen Chemie; dieser Band von Carl Graebe faßt nicht nur die Resultate der experimentellen und theoretischen Forschungen zusammen, sondern der Verfasser zeigt auch, in welcher Form jene Ergebnisse veröffentlicht wurden. So gebraucht er — heute eine unerschöpfliche Fundgrube — in seinen Darlegungen mit Vorliebe Ausdrücke und Bezeichnungen der Originalabhandlungen.
Walden kam von der physikalischen Chemie her. Er schreibt, er sei an der „Elektrifizierung" der organischen sogenannten Nichtelektrolyte beteiligt gewesen. In seiner Darstellung gelang es Walden, nicht nur die großen Linien und die Entstehung grundlegender Forschungen herauszuarbeiten, sondern er schildert auch möglichst vollständig die Objekte der chemischen Arbeit und die Forscher selbst unter genauer Angabe der Literatur.

Springer-Verlag Berlin · Heidelberg · New York

In kritischen Übersichten werden in dieser Reihe Stand und Entwicklung aktueller chemischer Forschungsgebiete beschrieben. Sie wendet sich an alle Chemiker in Forschung und Industrie, die am Fortschritt ihrer Wissenschaft teilhaben wollen.

In der Regel werden nur Beiträge veröffentlicht, die ausdrücklich angefordert worden sind. Schriftleitung und Herausgeber sind aber für ergänzende Anregungen und Hinweise jederzeit dankbar. Manuskripte können in den „Fortschritten der chemichen Forschung" in Deutsch oder Englisch veröffentlicht werden.

Jeder Band der Reihe ist einzeln käuflich.

This series presents critical reviews of the present position and future trends in modern chemical research. It is addressed to all research and industrial chemists who wish to keep abreast of advances in their subject.

As a rule, contributions are specially commissioned. The editors and publishers will, however, always be pleased to receive suggestions and supplementary information. Papers are accepted for "Topics in Current Chemistry" in either German or English.

Any volume of the series may be purchased separately.